VECTOR ANALYSIS 2023 GUIDE FOR BEGINNERS

PAERL RUDGARS

Copyright © 2023 Paerl Rudgars

All rights reserved.

INTRODUCTION

Welcome to "Vector Analysis 2023 Guide for Beginners," a comprehensive, step-by-step guide designed to provide a firm foundation for your understanding of vectors, their properties, their operations, and their essential role in various scientific disciplines. As we proceed through this guide, it's my sincere hope that you'll find clarity, insight, and, ultimately, a sense of mastery over what can be a complex but fascinating subject.

So, what exactly is vector analysis? If you're new to the field, you might be wondering why we need another kind of mathematics when the old, familiar numbers and equations seem perfectly fine for everyday calculations. But once we delve into the realms of physics, engineering, computer science, and more, we begin to see that there are structures in our world—vectors—that traditional scalar mathematics can't quite capture.

At its most basic level, a vector is a quantity with both magnitude and direction. This may seem simple, but it's a profound shift from the realm of scalars—quantities that have only magnitude. This added complexity opens up a world of new mathematical possibilities. The first part of this guide is dedicated to familiarizing you with these fundamental concepts.

Starting from an intuitive understanding of scalars and vectors, we'll delve into equal and null vectors and explore the basic operations such as scalar multiplication, and addition and subtraction of vectors. Throughout these sections, we'll also introduce you to the concept of the unit vector, the resolution of a vector, and much more.

Building upon this foundation, the next section of the guide deals with the product of vectors, an area where the unique properties of vectors truly come into their own. Here, we will discuss dot product and cross product, scalar and vector triple products, and even the intriguing notion of reciprocal vectors. These operations are not merely abstract mathematical concepts—they are tools that will allow you to understand and solve real-world problems.

After that, we move on to a thrilling aspect of vector analysis: differentiation. Much like in calculus, differentiation in vector analysis gives us insight into the rate of change, but this time, with a spatial dimension. We will cover topics like limit, derivative, derivative of a vector, and the derivative of sum and products of vectors. You will also be introduced to scalar and vector fields, and to the concept of a partial derivative and the ∇ operator.

As we delve further into vector calculus, the terms 'gradient', 'divergence', and 'curl' will become familiar friends. They are key to understanding the behavior of vector fields in various scenarios. Moreover, the Laplacian operator—a second-order differential operator—will introduce you to a new level of complexity and subtlety in the field.

The fifth part of this guide will lead you into the world of vector integration. Starting with line integrals, we'll extend our journey into surface integrals, and then move onto explore two of the most significant theorems in vector calculus—the Divergence theorem and Stoke's theorem. Each of these topics builds on the one before it, ensuring that your understanding deepens and solidifies with each new chapter.

Following this, we will guide you through the application of vectors in different coordinate systems. In particular, we'll focus on cylindrical and spherical coordinate systems, which are crucial for working with problems in three dimensions and for interpreting real-world scenarios in fields such as physics and engineering.

Finally, we will see the power of vectors in action. In the last part of the guide, we'll apply vector analysis to various disciplines, such as mechanics, electromagnetism, and fluid mechanics. These applications will not only reinforce your understanding of the concepts but also illustrate the wide-ranging applicability of vector analysis.

As with any journey, it's essential to come prepared. A familiarity with high-school level algebra, trigonometry, and basic calculus will be useful. But fear not if you feel a bit rusty—our aim here is to guide you gently but confidently into this new realm of mathematical understanding. The goal isn't just to master techniques and solve problems, but to build a robust conceptual understanding of vectors that will serve you well in whatever field you choose to apply them.

Remember, every master was once a beginner, and understanding, like any journey, begins with a single step. Here, in this guide, you're taking your first steps into a fascinating world where mathematics meets the structure of reality in surprising and beautiful ways. It's a journey I'm excited to be taking with you.

Let the adventure of vector analysis begin. Welcome aboard!

It's important to underscore why we've chosen 2023 as the key year for this edition of "Vector Analysis Guide for Beginners." As with any field of study, vector analysis is not static. It continues to grow and develop, influenced by advancements in various scientific and technological fields. This edition aims to incorporate those recent developments, providing you with the most up-to-date methods and applications in the field.

One significant development of the recent years has been the increased importance of vector analysis in machine learning and artificial intelligence. In these areas, vectors and vector operations are used to handle vast amounts of data and to create models that can 'learn' from this data. This guide, while firmly rooted in the fundamentals, is also forward-looking—it recognizes that the future of many disciplines, from physics to computer science to economics, will be increasingly data-driven. As such, understanding vector analysis will be a crucial tool in your mathematical toolbox.

Another feature of this edition is its accessibility. The content is carefully structured to make the principles of vector analysis approachable and digestible, regardless of your mathematical background. Each chapter gradually builds on the concepts introduced in the previous one, allowing you to grasp even the most complex aspects of vector analysis in a coherent and intuitive way. Clear, concise explanations are complemented by a variety of diagrams, illustrations, and practical examples, making the learning process dynamic and engaging.

In addition, this guide also features a variety of problem sets at the end of each chapter. These problems are designed to reinforce what you've learned and to challenge you to apply your new knowledge in novel ways. Solutions are provided, so you can check your understanding and clarify any points of confusion.

However, this guide isn't just about learning; it's also about discovery. As we journey through the world of vectors together, you'll encounter some of the most beautiful and elegant ideas in mathematics. You'll come to see how these ideas reflect the structure of the world around us and underpin much of our modern understanding of the universe. This, after all, is one of the great joys of studying mathematics—the thrill of discovery, the pleasure of understanding, and the profound appreciation for the beauty of the subject.

In this sense, this guide is not just a tool for learning—it's a gateway into a new way of seeing the world. As you turn the pages, you'll be embarking on a journey that's not just about acquiring knowledge, but about deepening your understanding of the universe and your place within it.

To conclude, the purpose of this guide is not only to make you proficient in the language of vector analysis but to foster a deep appreciation of its power, elegance, and applicability. By the end, my hope is that you'll not only have gained a solid understanding of vector analysis but that you'll also see the world through a new, mathematically-enlightened lens.

Here's to the journey that lies ahead. The world of vector analysis awaits, and I look forward to exploring it together with you in this guide, "Vector Analysis 2023 Guide for Beginners." Let's embark on this mathematical adventure!

CONTENTS

PREFACE 1

1. INTRODUCTION TO VECTORS 2
1.1. SCALARS AND VECTORS 2
1.2. EQUAL & NULL VECTORS 4
1.3. SCALAR MULTIPLICATION 5
1.4. ADDITION OF VECTORS 5
1.5. SUBTRACTION OF VECTORS 11
1.6. UNIT VECTOR 12
1.7. RESOLUTION OF VECTOR 13

2. PRODUCT OF VECTORS 19
2.1. DOT PRODUCT 19
2.2. CROSS PRODUCT 25
2.3. SCALAR TRIPLE PRODUCT 30
2.4. VECTOR TRIPLE PRODUCT 34
2.5. RECIPROCAL VECTORS 39

3. VECTOR DIFFERENTIATION 42
3.1. LIMIT 42
3.2. DERIVATIVE 44
3.3. DERIVATIVE OF A VECTOR 47
3.4. DERIVATIVE OF SUM & PRODUCTS OF VECTORS 51

3.5. SCALAR AND VECTOR FIELDS 53

3.6. PARTIAL DERIVATIVE & THE ∇ OPERATOR 56

4. GRADIENT, DIVERGENCE & CURL 59

4.1. GRADIENT 59

4.2. DIVERGENCE 65

4.3. CURL 69

4.4. MORE PROPERTIES 74

4.5. LAPLACIAN 79

5. VECTOR INTEGRATION 81

5.1. LINE INTEGRAL 81

5.2. SURFACE INTEGRAL 97

5.3. DIVERGENCE THEOREM 111

5.4. STOKE'S THEOREM 115

6. CURVILINEAR COORDINATE SYSTEMS 120

6.1. CYLINDRICAL COORDINATE SYSTEM 121

6.2. SPHERICAL COORDINATE SYSTEM 136

7. APPLICATIONS 150

7.1. MECHANICS 150

7.2. ELECTROMAGNETISM 165

7.3. FLUID MECHANICS 174

APPENDIX **180**

PREFACE

Vector analysis is a very useful and a powerful tool for physicists and engineers alike. It has applications in multiple fields. Although it is not a particularly difficult subject to learn, students often lack a proper understanding of the concepts on a deeper level. This restricts its usage to a mere mathematical tool.

That's where this book hope to be different. We don't want this subject to be treated just as a mathematical tool. We hope to go beyond it. Therefore, the emphasis is to provide physical interpretation to the various concepts in the subject with the help of illustrative figures and intuitive reasoning. Having said that, we have given adequate importance to the mathematical aspect of the subject as well. 100+ solved examples given in the book will give the reader a definite edge when it comes to problem solving.

For beginners this book will provide a concise introduction to the world of vectors in a unique way. The various concepts of the subject are arranged logically and explained in a simple reader-friendly language, so that they can learn with minimum effort in quick time. For experts, this book will serve as a good refresher.

The first 2 chapters focus on the basics of vectors. In chapters 3 to 5 we dig into vector calculus. Chapter 6 is all about vectors in different coordinate systems and finally chapter 7 focuses on the applications of vectors in various fields like engineering mechanics, electromagnetism, fluid mechanics etc.

Readers are welcome to give constructive suggestions for the improvement of the book.

Enjoy this wonderful subject. Thank you!

1. INTRODUCTION TO VECTORS

1.1 SCALARS AND VECTORS

In physics (and mathematics), quantities can be classified into 2 main categories; Scalars and Vectors. Certain quantities such as mass, time, temperature, volume, density etc. can be adequately represented just by a numerical value, an amount or a magnitude. Such quantities are called Scalars. For example, if the mass of an object is said to be 2 kilograms, the value 2 denotes the number of times the unit kilogram is contained in that object.

On the other hand, certain other quantities such as force, velocity, displacement, acceleration etc. cannot be adequately represented by a numerical value alone. The only way to meaningfully represent such quantities is by considering them as having a direction in addition to the magnitude. Such quantities that have a magnitude as well as a direction are called Vectors. For example, if you walk 2 miles north, then walk 2 miles south, you are back at your original position i.e. your net displacement is zero. In this case, had we just mentioned the magnitude and ignored the directions, there's no way of telling your net displacement, you could have walked 4 miles straight or 4 miles left or any which way.

Vectors are represented by a directed line segment, basically arrows. The length of the line segment denotes the magnitude and the arrowhead indicates the direction of the vector. For example, a vector from point A to another point B is denoted by $\overrightarrow{AB}$. Point A is called the tail of the vector and point B is called the head of the vector. The magnitude of a vector $\overrightarrow{AB}$ is denoted as $|AB|$.

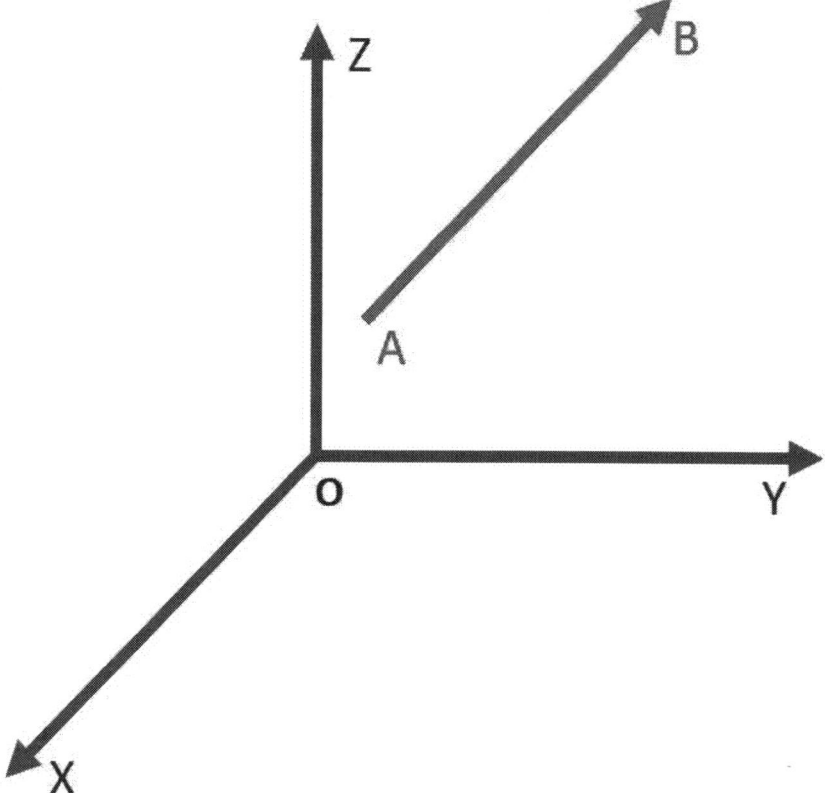

Note that a vector is characterized only by its magnitude and direction, and not by its position in space. What this means is that, as long as the magnitude and the direction are unaltered, you are free to move the vector around.

Solved Examples:

1. Classify these quantities as a scalar or a vector.

a) 5m
b) 10 km's North
c) 3000 Calories
d) 60 km/hr East

Option **a** and **c** are scalar quantities as they do not have a direction indicated for them. Option **b** and **c** are vector quantities as they have both a magnitude and a direction.

2. Calculate the distance travelled and net displacement in each of the following cases.

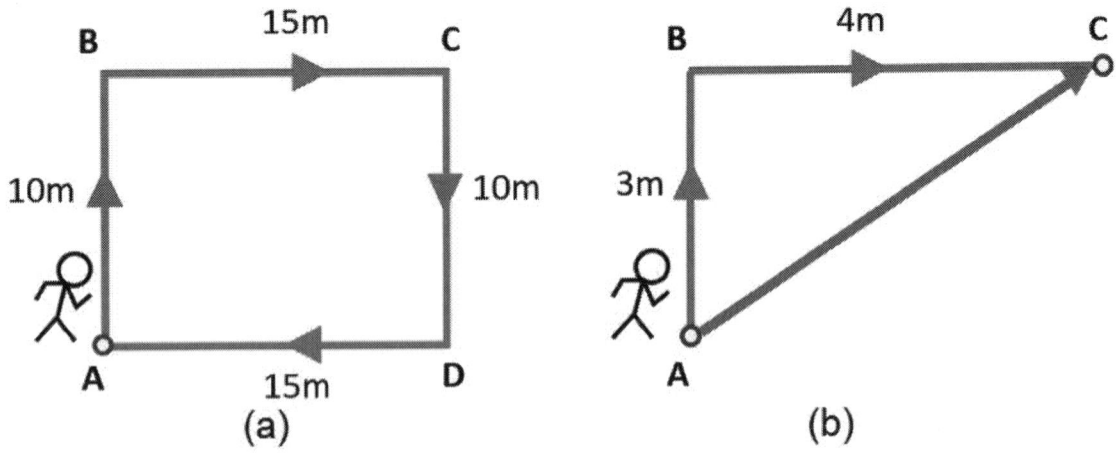

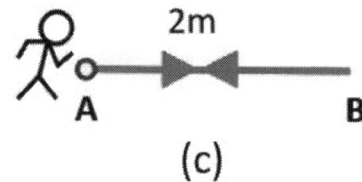

Distance is a scalar quantity, whereas Displacement is a vector quantity. Displacement solely depends on the starting and the end positions, independent of the path taken.

In case (a), distance = ABCD = 50m, displacement = $\vec{AA}$ = 0

In case (b), distance = ABC = 7m, displacement = $\vec{AC}$ = $\sqrt{3^2 + 4^2}$ = 5m

In case (c), distance = AB + BA = 4m, displacement = $\vec{AA}$ = 0

*From this point on, we'll denote vectors in capital letters without the arrow superscript.

1.2 EQUAL & NULL VECTORS

Two vectors are said to be equal if they have the same magnitude and the same direction. The equality of two vectors A and B is denoted as,

$$\vec{A} = \vec{B}$$

Again, their spatial location is irrelevant, it's their magnitude and direction that matters.

A vector with zero magnitude is called as a Null Vector or a zero vector. In case of null vectors, the notion of direction can be disregarded. For this reason, all null vectors are equal.

1.3 SCALAR MULTIPLICATION

When a vector is multiplied by a positive scalar quantity (a positive number), its magnitude gets multiplied by the scalar quantity and the direction remains unaltered. Similarly, if a vector is multiplied by a negative scalar quantity, its magnitude gets multiplied by the scalar quantity and its direction reverses.

The product of a scalar **x** and vector A is denoted as **x**A.

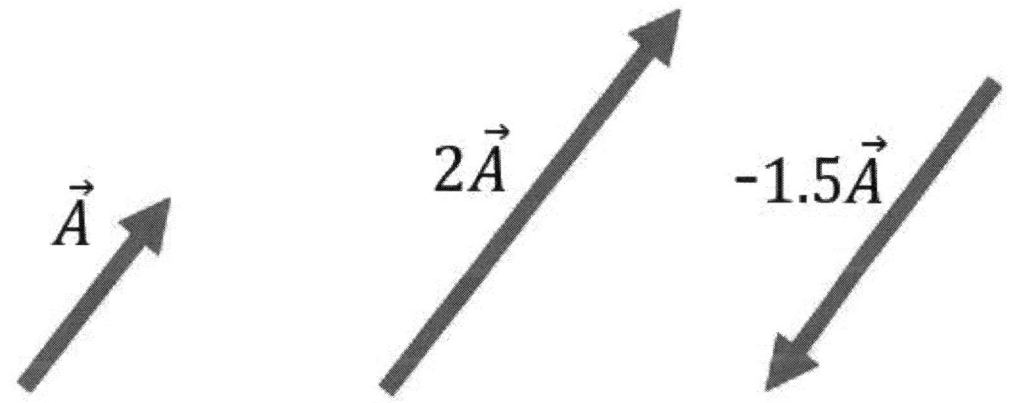

1.4 ADDITION OF VECTORS

Vector addition isn't as straightforward as adding scalar quantities. In case of vectors, the directions have to be considered as well. Consider a solid block being applied 10 Newtons of force from 2 directions as shown in the figure below.

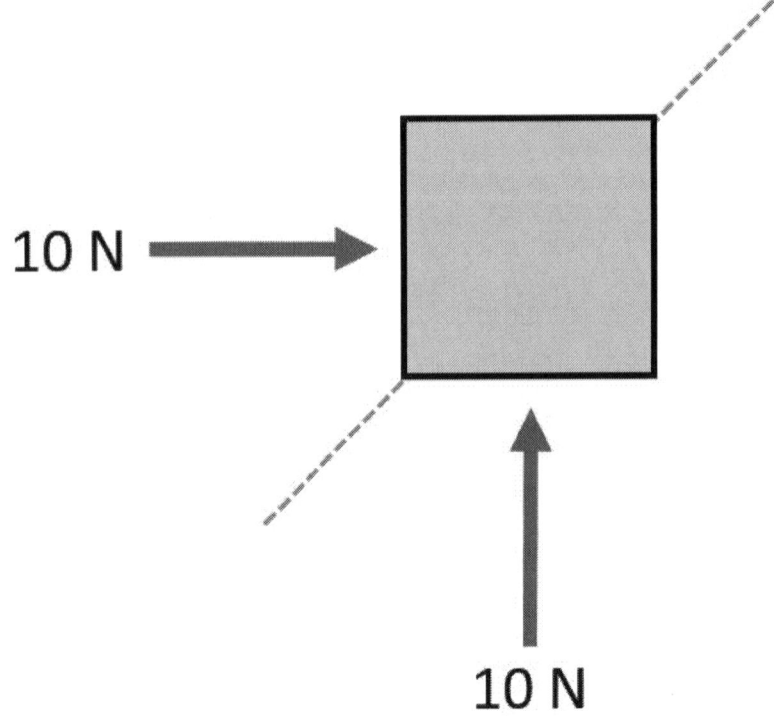

In which direction do you think the block will move as a result of these 2 forces? Of course, along the dotted line. Now imagine that the block is being pushed with 20 Newtons of force from the bottom and 10 Newtons of force from the side as before. In which direction do you think the block will move now? Common sense says the block will move in a direction slightly to the left of the dotted line.

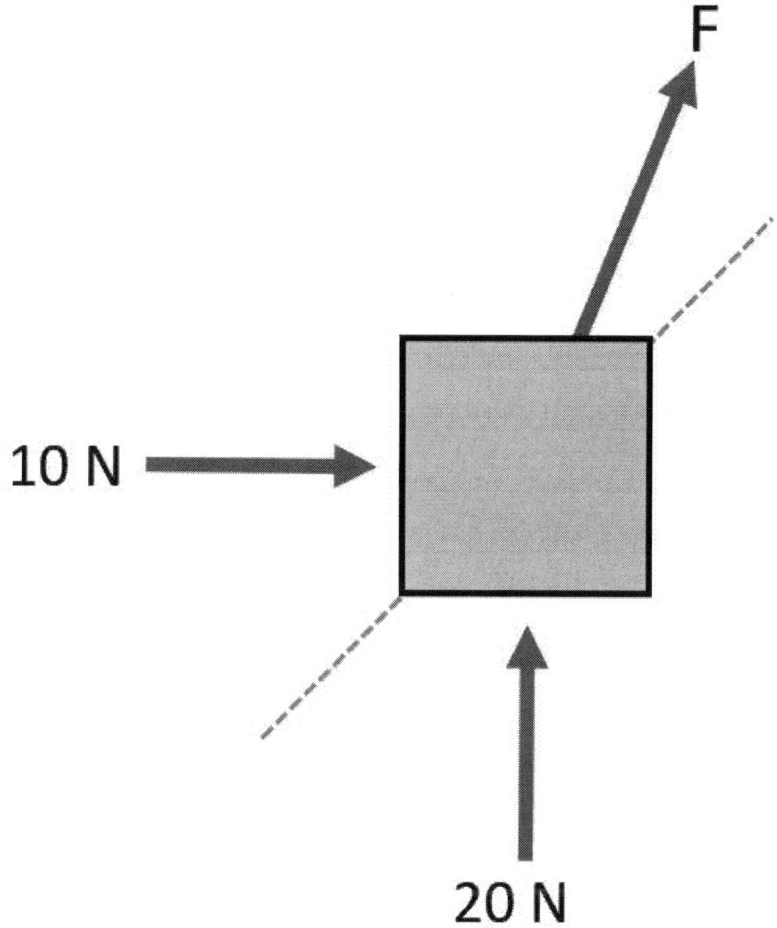

What we just did with these 2 examples is Vector addition. In general, 2 vector quantities can be added together using the **Triangle law of Vector Addition**. It states that *when two vectors are represented as two sides of a triangle, then the third side of the triangle represents the resultant vector.*

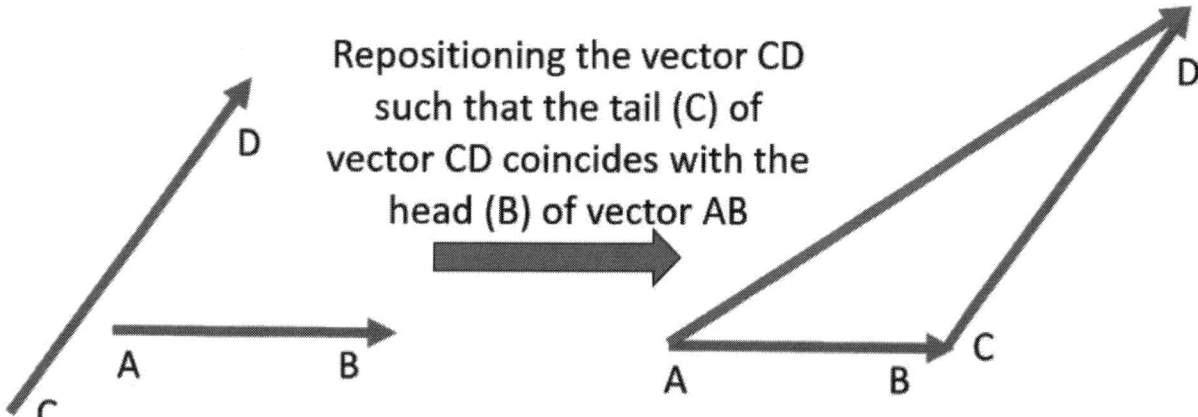

In the figure above, we repositioned vector CD such that the tail (C) of vector CD coincides with the head (B) of vector AB. Hence the resultant is given by the vector AD. Now if you try the same, but by repositioning vector AB instead of CD, you will get the exact same result, proving that vector addition is commutative.

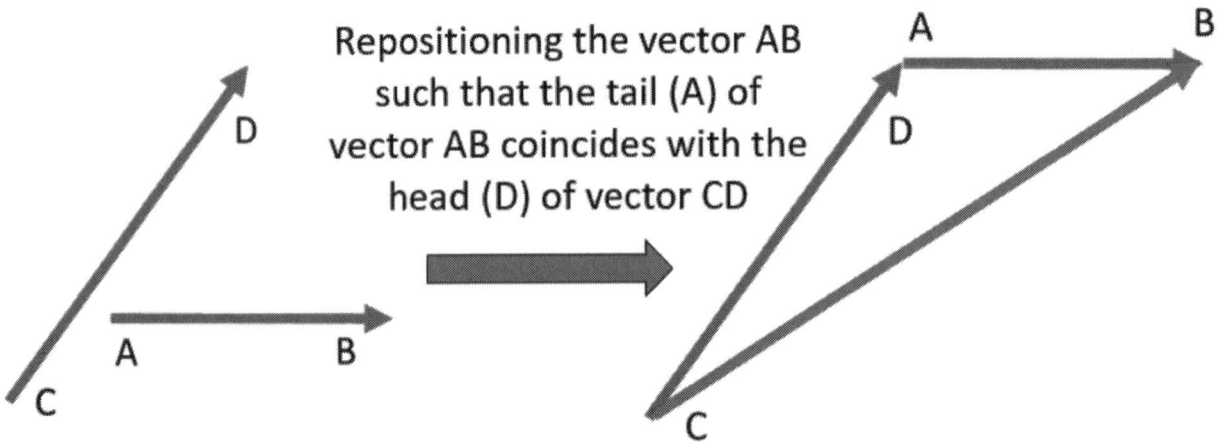

The magnitude and the direction of the resultant vector is given by,

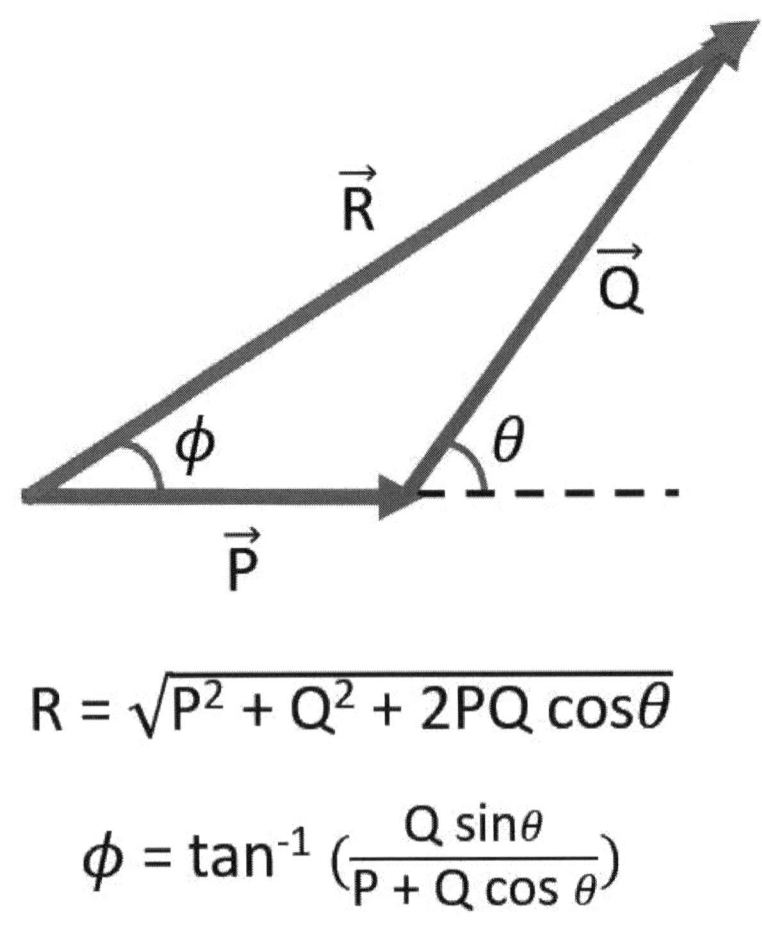

$$R = \sqrt{P^2 + Q^2 + 2PQ\cos\theta}$$

$$\phi = \tan^{-1}\left(\frac{Q\sin\theta}{P + Q\cos\theta}\right)$$

Check the Appendix for the proof.

Note that ϕ is the angle made by the resultant with respect to the vector P. To find the angle made by the resultant with respect to vector Q, switch P and Q in the above formula.

We can extend this idea of triangle law to any no. of vectors and the result is the **Polygon law of vector addition**.

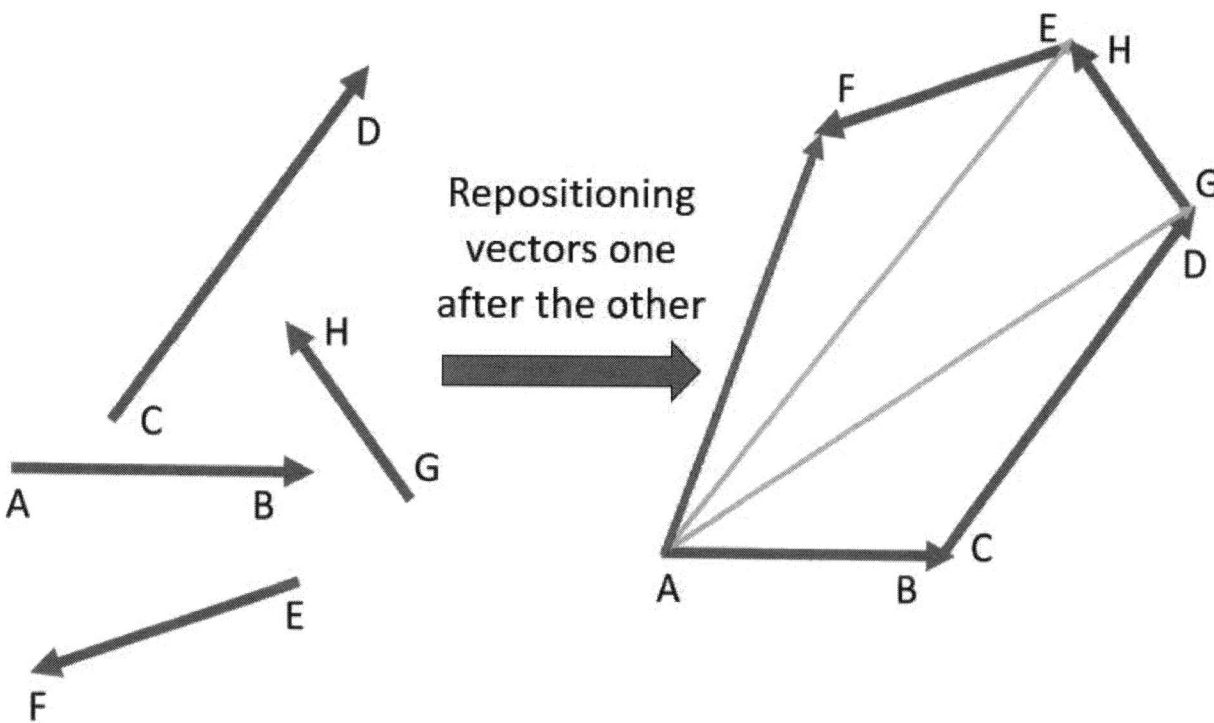

Properties of Vector Addition:

1. **Associative Property:** (A + B) + C = A + (B + C)
2. **Commutative Property:** A + B = B + A
3. **Distributive Property:** k (A + B) = k A + k B
4. (m + n) A = m A + Na
5. A + 0 = 0 + A = A
6. A + (-A) = 0

Where **k, m, n** are scalars.

Solved Examples:

3. **If 2 forces of magnitude 4N and 3N act on a solid block as shown in the figure below, Calculate the net force F acting on the block.**

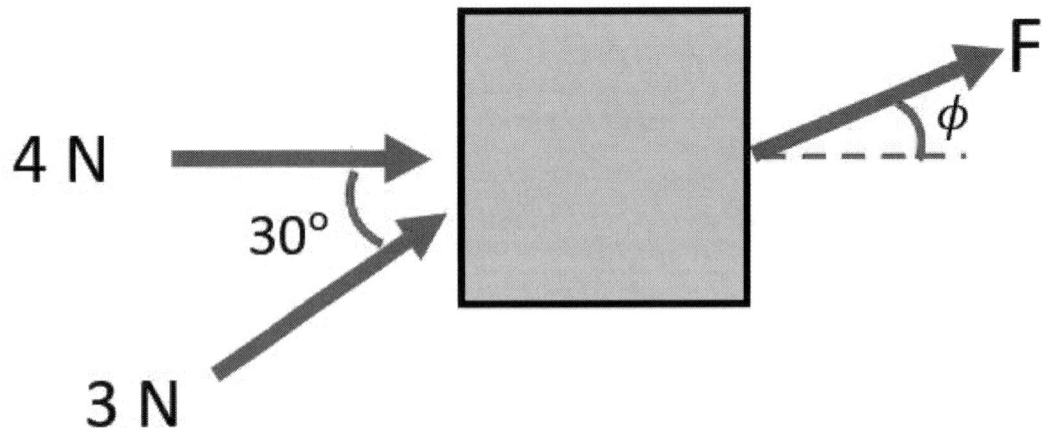

$$F = \sqrt{F_1^2 + F_2^2 + 2F_1F_2 \cos\theta}$$
$$= \sqrt{4^2 + 3^2 + 2(4)(3)\cos 30}$$
$$= \sqrt{25 + 24(\tfrac{\sqrt{3}}{2})}$$
$$= \underline{\underline{6.766 \text{ N}}}$$

$$\phi = \tan^{-1}\left(\frac{F_2 \sin\theta}{F_1 + F_2 \cos\theta}\right)$$
$$= \tan^{-1}\left(\frac{3 \sin 30}{4 + 3\cos 30}\right)$$
$$= \underline{\underline{12.8°}}$$

4. A cyclist heads north with a velocity of 40 km/hr. A cross wind blows from East to West at 10 km/hr. Calculate the resultant velocity.

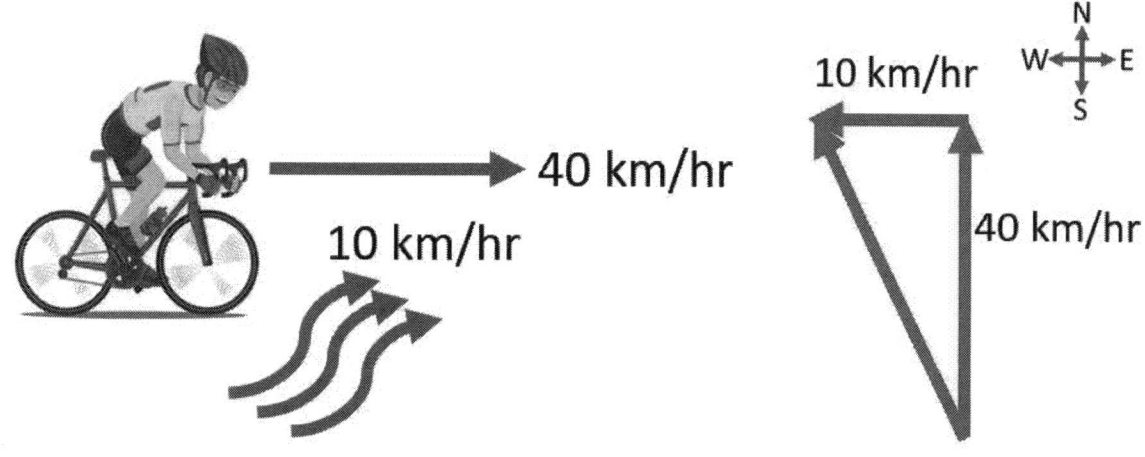

$$v_{resultant} = \sqrt{v_1^2 + v_2^2 + 2v_1v_2 \cos 90}$$
$$= \sqrt{40^2 + 10^2}$$
$$= 41.23 \text{ km/hr}$$

1.5 SUBTRACTION OF VECTORS

Subtracting a vector Q from another vector P is basically the same as adding vector P and the reverse of vector Q i.e. **P-Q = P + (-Q)**.

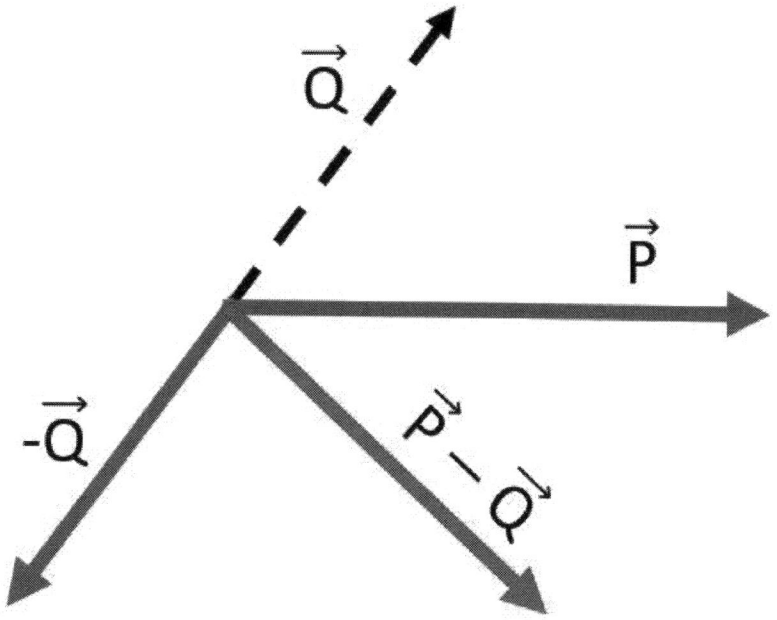

The magnitude of the resultant correspondingly becomes,

$$R = \sqrt{P^2 + Q^2 - 2PQ \cos\theta}$$

Solved Examples:

5. **The resultant of 2 forces P = 30N and Q is R = 40N, inclined at 60° to P. Find the magnitude of Q.**

Force R is the resultant of forces P and Q, therefore Q can be expressed as the difference of R and P.

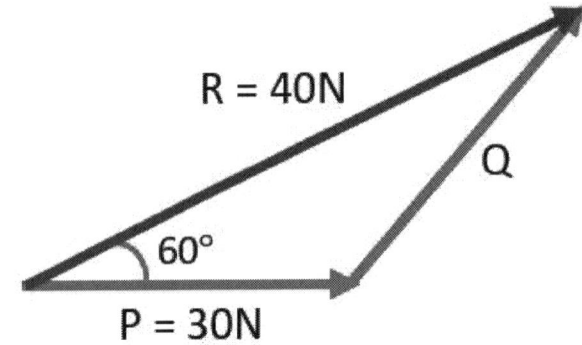

$$Q = \sqrt{P^2 + R^2 - 2PR \cos\theta}$$
$$= \sqrt{30^2 + 40^2 - 2(30)(40) \cos 60}$$
$$= 36.05 \text{ N}$$

1.6 UNIT VECTOR

Along a particular direction there can be infinite no. of possible vectors, with all of them differing only in magnitudes from one another. But in any direction, there can only be a single vector of unit magnitude, such a vector is called the Unit vector. The significance of the Unit vector is that all vectors in a specific direction are scaled versions of the unit vector in that direction.

A unit vector in the direction of a vector $\vec{A}$ is denoted by $\hat{A}$. The unit vector along any vector can be obtained by dividing the vector by its magnitude as,

$$\widehat{A} = \frac{\vec{A}}{|A|}$$

1.7 RESOLUTION OF VECTOR

One big advantage of using vectors is that they can be resolved into any no. of components. In the simplest case, a vector can be resolved into 2 component vectors lying in the same plane.

Shown below is a vector OA, if we draw its projections onto any 2 perpendicular axes in its plane (for convenience we have picked the x and y axes), we get a set of 2 new vectors OU and OV. These 2 new vectors are called the components of vector OA.

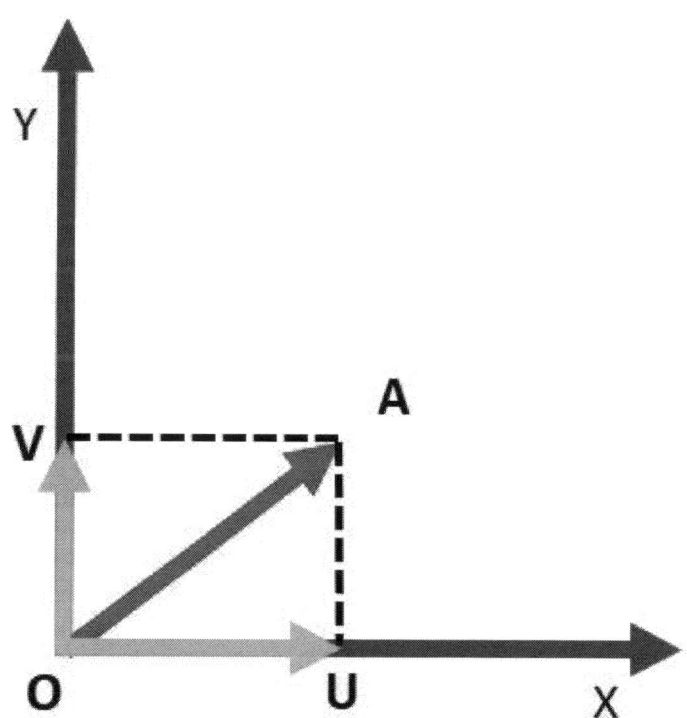

We can take this one step further and resolve a vector into 3 components in 3 dimension. What's the advantage of doing this? For one, resolving vectors into perpendicular components allows us to express them in terms of the Cartesian coordinates, which we are familiar with.

Consider a vector OB as shown in the figure below. If we resolve this vector into 2 perpendicular vectors, we get 2 component vectors OG and OV. Vector OG is along the z- axis, so we'll leave it as such. But vector OV on the other hand lies in the x-y plane, so we consider OV as a separate vector and resolve it further. And we get its 2 component vectors OE and OF along the x and y axes respectively. This way we have resolved the vector OB into 3 mutually perpendicular vectors OE, OF and OG, which lies along the cartesian coordinates.

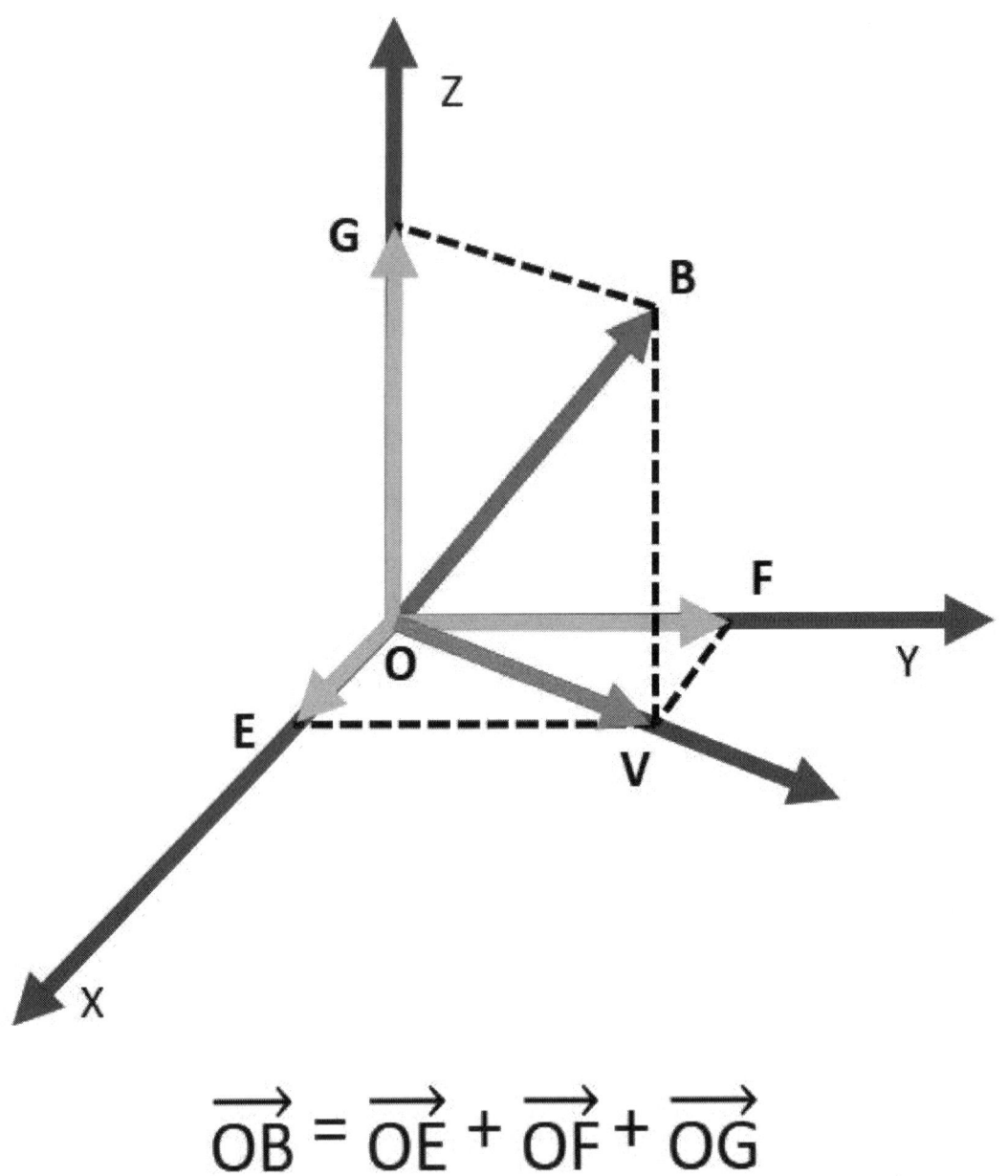

$$\vec{OB} = \vec{OE} + \vec{OF} + \vec{OG}$$

Any vector can be resolved into components along the coordinate axes in this manner. By doing this, we have managed to generalize vectors a little bit. Using the idea of unit vectors from last section we can generalize vectors even more.

As mentioned in the previous section, the basic idea behind the unit vector is that it is the most basic vector along a direction and every other vector along that same direction is just a scaled version of that unit vector. So if we denote the unit vectors along x, y and z axes as $\hat{i}, \hat{j}$ and $\hat{k}$ (Engineering textbooks might use $\hat{a}_x, \hat{a}_y$ and $\hat{a}_z$ notation) respectively, then any vector A can be expressed as,

$$\vec{A} = \text{(Magnitude in x direction)} \times \hat{i}$$
$$+ \text{(Magnitude in y direction)} \times \hat{j}$$
$$+ \text{(Magnitude in z direction)} \times \hat{k}$$

As a consequence of this generalization, we can denote a vector simply by using 3 no's. For example, (5, 2, -4) denotes the vector $5\hat{i} + 2\hat{j} - 4\hat{k}$. Similarly, (0, 1, 2) denotes the vector $\hat{j} + 2\hat{k}$. In the later example the x component is absent, meaning the vector lies in the y-z plane.

The magnitude of a vector can be obtained from its components as,

$$\text{Magnitude of a vector} = \sqrt{\text{(Magnitude in x direction)}^2 + \text{(Magnitude in y direction)}^2 + \text{(Magnitude in z direction)}^2}$$

This result can be easily obtained by using the Pythagoras theorem.

In general any vector A can be expressed in terms of any three non-coplanar vectors **a**, **b** and **c**, which are not necessarily unit vectors.

$$\vec{A} = p\vec{a} + q\vec{b} + r\vec{c}$$

Where **p, q** and **r** are 3 scalars.

Solved Examples:

6. If vector A = $3\hat{i} + 2\hat{j} + 2\hat{k}$. Find the vector 5A - $3\hat{k}$.

$$\vec{A} = 3\hat{i} + 2\hat{j} + 2\hat{k}$$
$$\Rightarrow 5\vec{A} = 15\hat{i} + 10\hat{j} + 10\hat{k}$$
$$\therefore 5\vec{A} - 3\hat{k} = 15\hat{i} + 10\hat{j} + 10\hat{k} - 3\hat{k}$$
$$= 15\hat{i} + 10\hat{j} + 7\hat{k}$$

7. Calculate the magnitude of the vector A = $5\hat{i} + 2\hat{j} - 4\hat{k}$.

$$\vec{A} = 5\hat{i} + 2\hat{j} - 4\hat{k}$$
$$\therefore |\vec{A}| = \sqrt{5^2 + 2^2 + (-4)^2}$$
$$= 3\sqrt{5}$$

8. Find the Unit vector along the direction of vector A = $2\hat{i} - 4\hat{j} + \hat{k}$.

$$\text{Unit Vector } \hat{A} = \frac{\vec{A}}{|\vec{A}|}$$

$$= \frac{2\hat{i} - 4\hat{j} + \hat{k}}{\sqrt{2^2 + (-4)^2 + 1^2}}$$

$$= \frac{2\hat{i} - 4\hat{j} + \hat{k}}{\sqrt{21}}$$

$$= \frac{2}{\sqrt{21}}\hat{i} - \frac{4}{\sqrt{21}}\hat{j} + \frac{1}{\sqrt{21}}\hat{k}$$

9. Obtain a vector of magnitude 10 in the direction of the vector A = $5\hat{i} + 3\hat{j} + \hat{k}$

The Unit vector is the most basic vector in a particular direction, therefore by scaling the unit vector by the appropriate magnitude we can obtain the required result.

$$\text{Unit Vector } \hat{A} = \frac{\vec{A}}{|\vec{A}|}$$

$$= \frac{5\hat{i} + 3\hat{j} + \hat{k}}{\sqrt{5^2 + 3^2 + 1^2}}$$

$$= \frac{5\hat{i} + 3\hat{j} + \hat{k}}{\sqrt{35}}$$

$$\therefore 10\hat{A} = 10\left(\frac{5\hat{i} + 3\hat{j} + \hat{k}}{\sqrt{35}}\right)$$

$$= \frac{50}{\sqrt{35}}\hat{i} + \frac{30}{\sqrt{35}}\hat{j} + \frac{10}{\sqrt{35}}\hat{k}$$

10.
  A person pushes his lawnmower with 60N of force at angle of 60° with the ground. Find the horizontal and vertical components of the applied force.

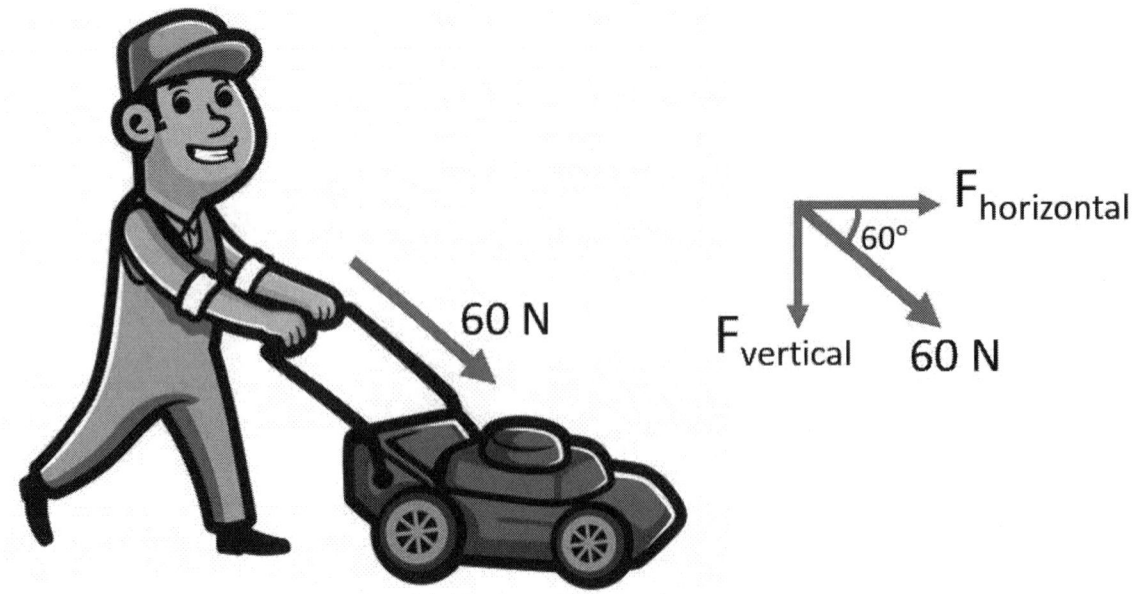

Rearranging the forces into a right triangle, the components can be obtained as,

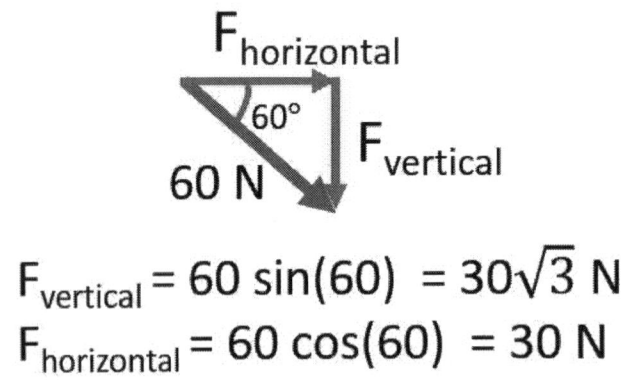

$$F_{vertical} = 60 \sin(60) = 30\sqrt{3} \text{ N}$$
$$F_{horizontal} = 60 \cos(60) = 30 \text{ N}$$

If you take the resultant of the vertical and horizontal components obtained, you'll get the original force 60N, hence verifying our result.

2. PRODUCT OF VECTORS

2.1 DOT PRODUCT

Vectors can be multiplied in 2 possible ways; the Dot product and the Cross product. The dot product of 2 vectors results in a scalar quantity and the cross product of 2 vectors results in another vector. Hence, the dot product and the cross product are also known as the scalar product and the vector product respectively.

The dot product between 2 vectors is denoted as **A.B** (read as A dot B). It can be obtained as,

$$\vec{A} \cdot \vec{B} = |A||B| \cos \theta$$

$|A|$ = Magnitude of $\vec{A}$
$|B|$ = Magnitude of $\vec{B}$
$\cos \theta$ = cos of the angle between $\vec{A}$ and $\vec{B}$

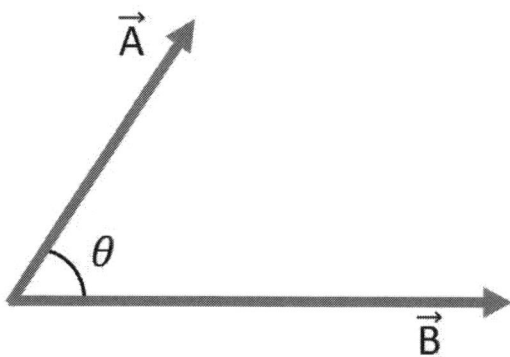

Intuitively, the dot product can be thought of as a measure of similarity of two vectors or how well they work together with one another. Consider our "forces on a block" example again.

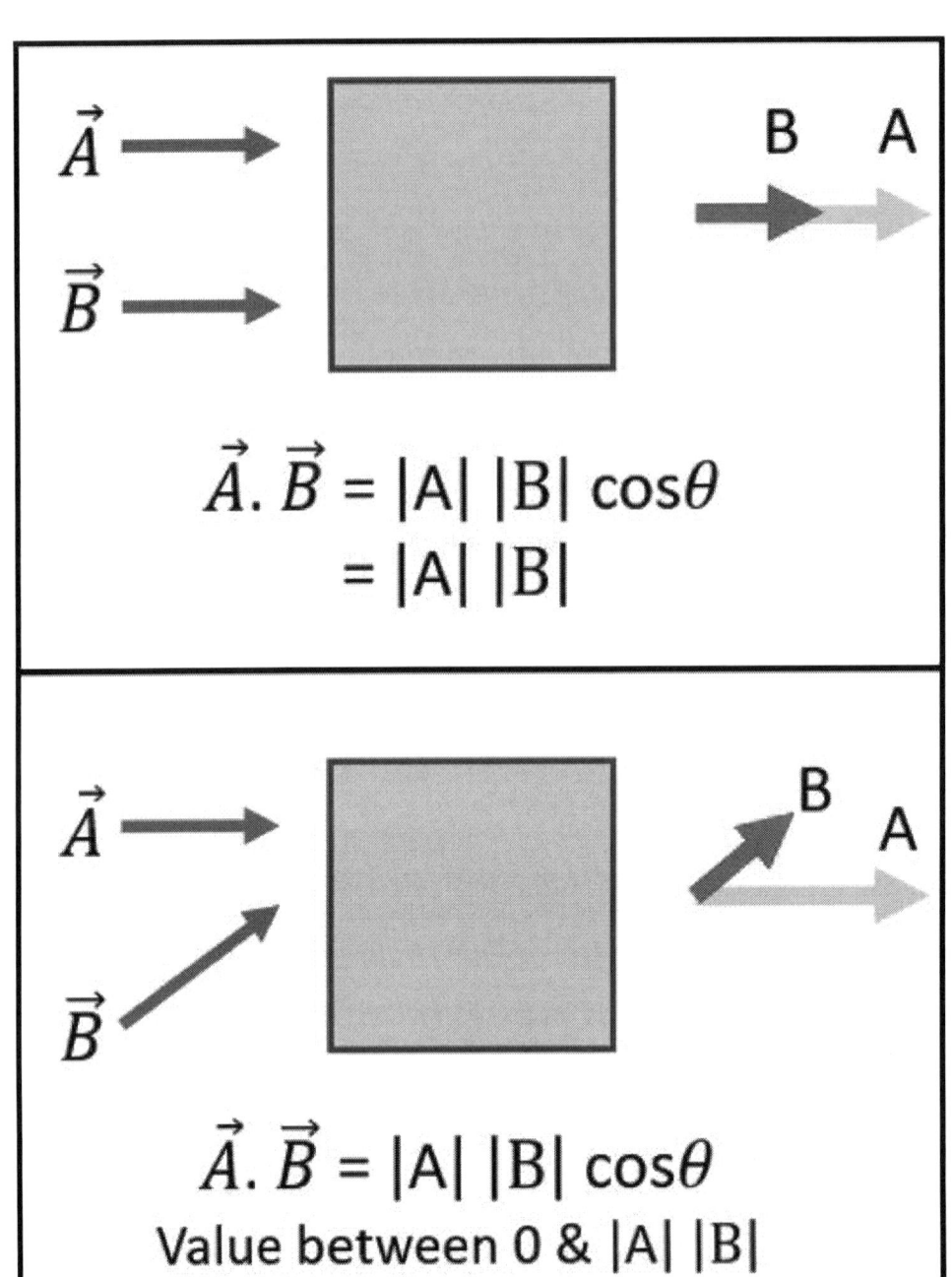

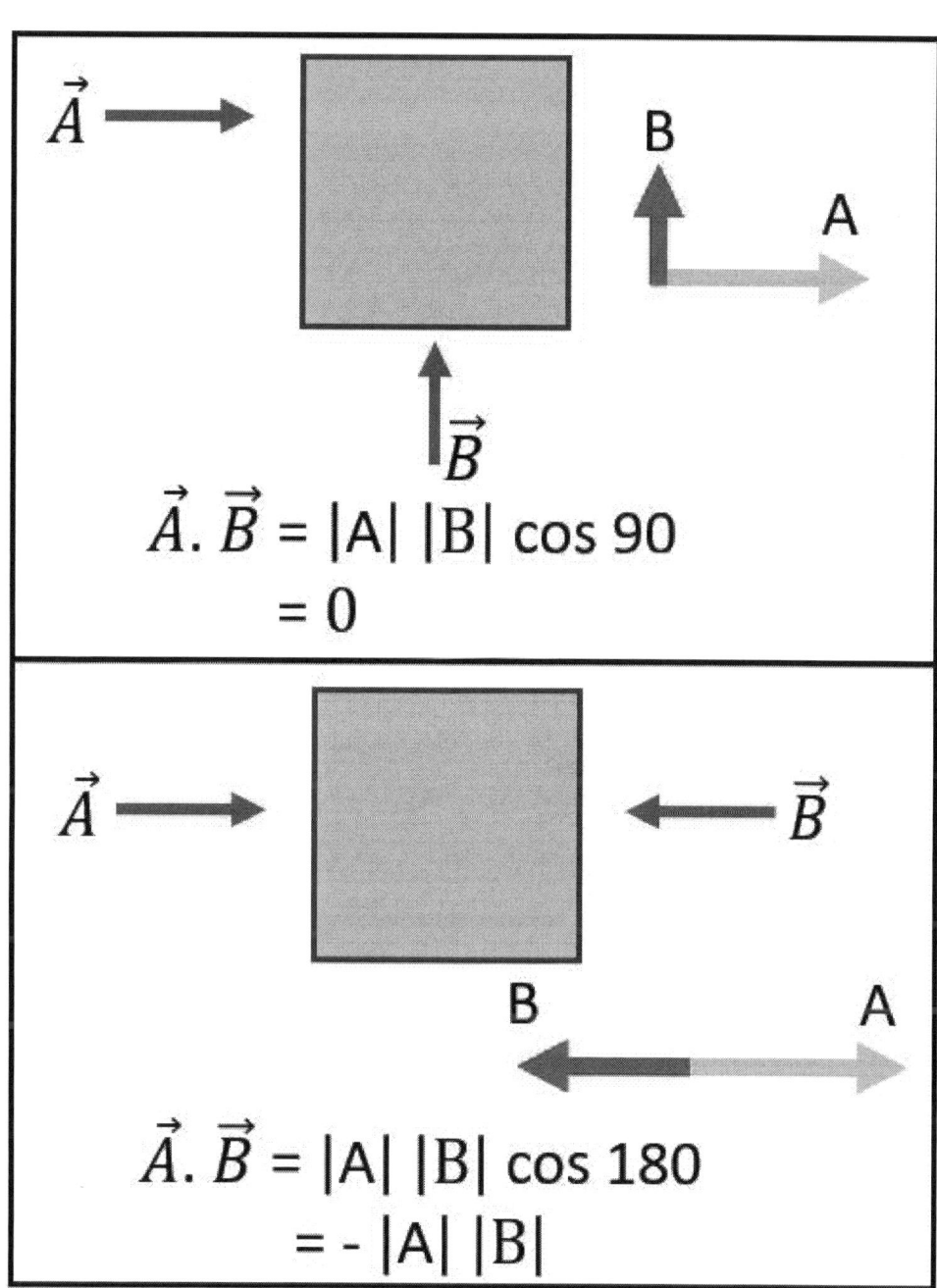

In the first case, the forces A and B are in the same direction and hence they are working together to move the block. Therefore, the dot product is maximum in this case. In the second case, the force B is applied at an angle θ to the force A, this obviously is not the best method to move the block. The force B is not contributing to the motion of the block as much as it did in the first case, hence the dot product is a smaller value, but non zero. In the third case, the force B is applied orthogonally to the force A, so it doesn't contribute to moving the block (in the direction of A) at all, hence the dot product here is zero. In the fourth case, the force B is applied in opposite direction to force A, which means force B is not only not contributing to the motion of the block, but it's actually negating the effect of force A. So, the dot product in this case is negative.

Geometrically, the dot product can be interpreted in terms of projection. If **a** and **b** are two unit then **a.b** gives the length of the projection of either vector on the other.

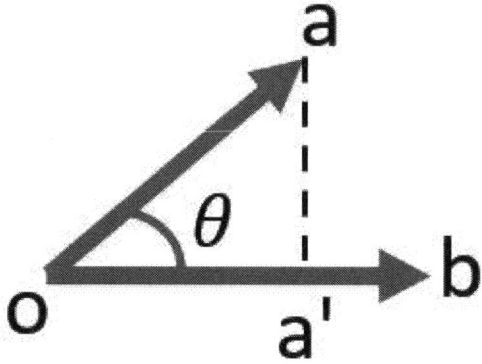

$$\vec{a}.\vec{b} = |a| |b| \cos \theta$$
$$= oa \times ob \times \cos \theta$$

In Δoaa',

$$\cos \theta = \frac{oa'}{oa}$$

$$\therefore \vec{a}.\vec{b} = oa \times ob \times \frac{oa'}{oa}$$
$$= oa' \ (\because oa = ob = 1)$$

In general, if A and B are any two vectors, then A.B gives the product of the length of that vector and the length of its projection upon the other.

If the angle between two vectors is 90°, then the dot product vanishes. This property is used to check the perpendicularity of 2 vectors. Similarly, if the angle between the vectors is 0°, the dot product is simply the product of their magnitudes. Consequently, the dot products of the 3 fundamental unit vectors are,

$$i.j = j.k = k.i = 0$$
$$i.i = j.j = k.k = 1$$

Properties of Dot Product:

- **Commutative property:** A.B = B.A
- **Distributive property:** A.(B + C) = A.B + A.C
- (A + B).(C + D) = A.C + A.D + B.C + B.D
- k (A.B) = (kA).B = A.(kB), where **k** is a scalar
- A.A = $|A|_2$

Solved Examples:

1. Obtain the dot product of vectors A = $\hat{i} + 2\hat{j} - 3\hat{k}$ and B = $3\hat{i} + 5\hat{j} + 7\hat{k}$.

$$\vec{A} = \hat{i} + 2\hat{j} - 3\hat{k}, \vec{B} = 3\hat{i} + 5\hat{j} + 7\hat{k}$$
$$\vec{A}.\vec{B} = (\hat{i} + 2\hat{j} - 3\hat{k}).(3\hat{i} + 5\hat{j} + 7\hat{k})$$
$$= (\hat{i}.3\hat{i} + 2\hat{j}.5\hat{j} - 3\hat{k}.7\hat{k})$$
$$= 3 + 10 - 21$$
$$= -8$$

2. Find the value of 'a' for which vectors A = $2\hat{i} + \hat{j} + 5\hat{k}$ and B = $a\hat{i} - 3\hat{j} + 4\hat{k}$ are perpendicular to each other.

$$\vec{A} = 2\hat{i} + \hat{j} + 5\hat{k},\ \vec{B} = a\hat{i} - 3\hat{j} + 4\hat{k}$$

$$\vec{A}.\vec{B} = (2\hat{i} + \hat{j} + 5\hat{k}).(a\hat{i} - 3\hat{j} + 4\hat{k})$$
$$= 2a - 3 + 20 = 0$$
$$\Rightarrow a = -8.5$$

3. Find the angle between the vectors $A = 2\hat{i} + \hat{j} + 5\hat{k}$ and $B = \hat{i} + \hat{j} + \hat{k}$.

$$\vec{A} = 2\hat{i} + \hat{j} + 5\hat{k},\ \vec{B} = \hat{i} + \hat{j} + \hat{k}$$

$$\vec{A}.\vec{B} = (2\hat{i} + \hat{j} + 5\hat{k}).(\hat{i} + \hat{j} + \hat{k})$$
$$= 2 + 1 + 5 = 8$$

$$|\vec{A}| = \sqrt{2^2 + 1^2 + 5^2} = \sqrt{30}$$
$$|\vec{B}| = \sqrt{1^2 + 1^2 + 1^2} = \sqrt{3}$$

$$\cos\theta = \frac{\vec{A}.\vec{B}}{|\vec{A}||\vec{B}|} = \frac{8}{\sqrt{3}\sqrt{30}} = 0.8432$$

$$\therefore \theta = \cos^{-1}(0.8432) = 32.5°$$

2.2 CROSS PRODUCT

The second way of multiplying 2 vectors is called the Cross product and the cross product of two vectors results in another vector.

The cross product of 2 vectors A and B is denoted as **A x B**. It can be obtained as,

$$\vec{A} \times \vec{B} = |A||B| \sin \theta \, \hat{n}$$

$|A|$ = Magnitude of $\vec{A}$
$|B|$ = Magnitude of $\vec{B}$
$\sin \theta$ = sin of the angle between $\vec{A}$ and $\vec{B}$
$\hat{n}$ = Unit vector perpendicular to both the vectors

Interpreting the physical meaning of the cross product isn't as straightforward as with the dot product. Geometrically, the cross product of the two vectors represents the area of the parallelogram formed with these 2 vectors taken as adjacent sides. And the direction of the cross product vector is normal to plane of this parallelogram. In some sense the cross product can be thought of as a measure of the orthogonality of 2 vectors. Closer the angle between the vectors is to 90 degrees, larger the area of the parallelogram formed and correspondingly larger the magnitude of the cross product vector.

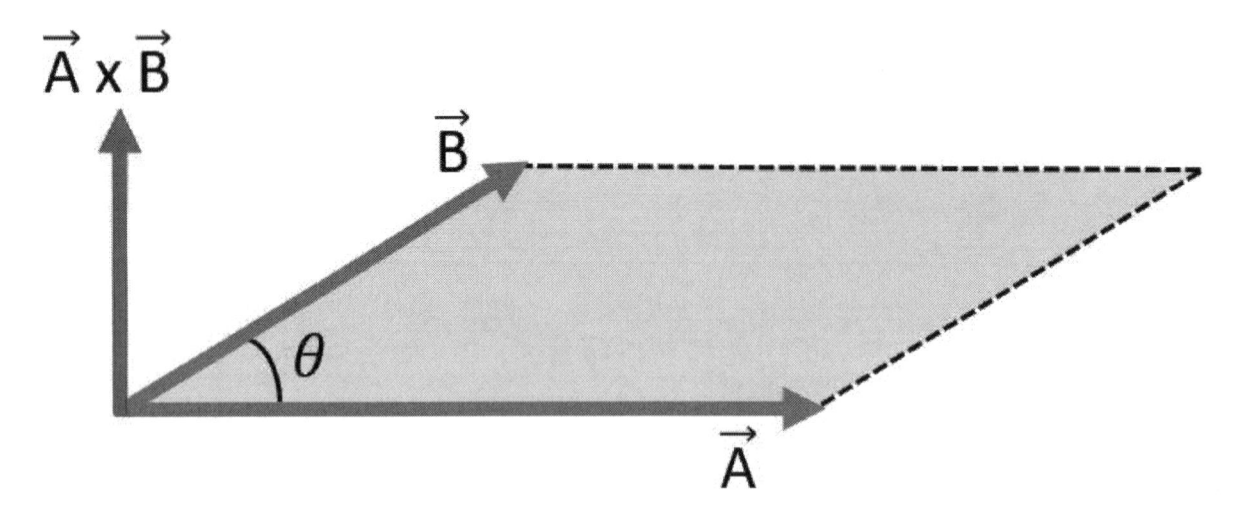

Now there's another small problem, given 2 vectors A and B, there are two possible directions that the cross product vector could point to, upwards and downwards. This is where the right hand thumb rule comes in, which states that *"If you curl your fingers of your right hand in such a way that the index finger*

points in the direction of vector A and middle finger points in the direction of vector B, then the thumb points in the direction of AxB". Now if you try the same for BxA, the thumb will point in the downward direction i.e. BxA = -AxB.

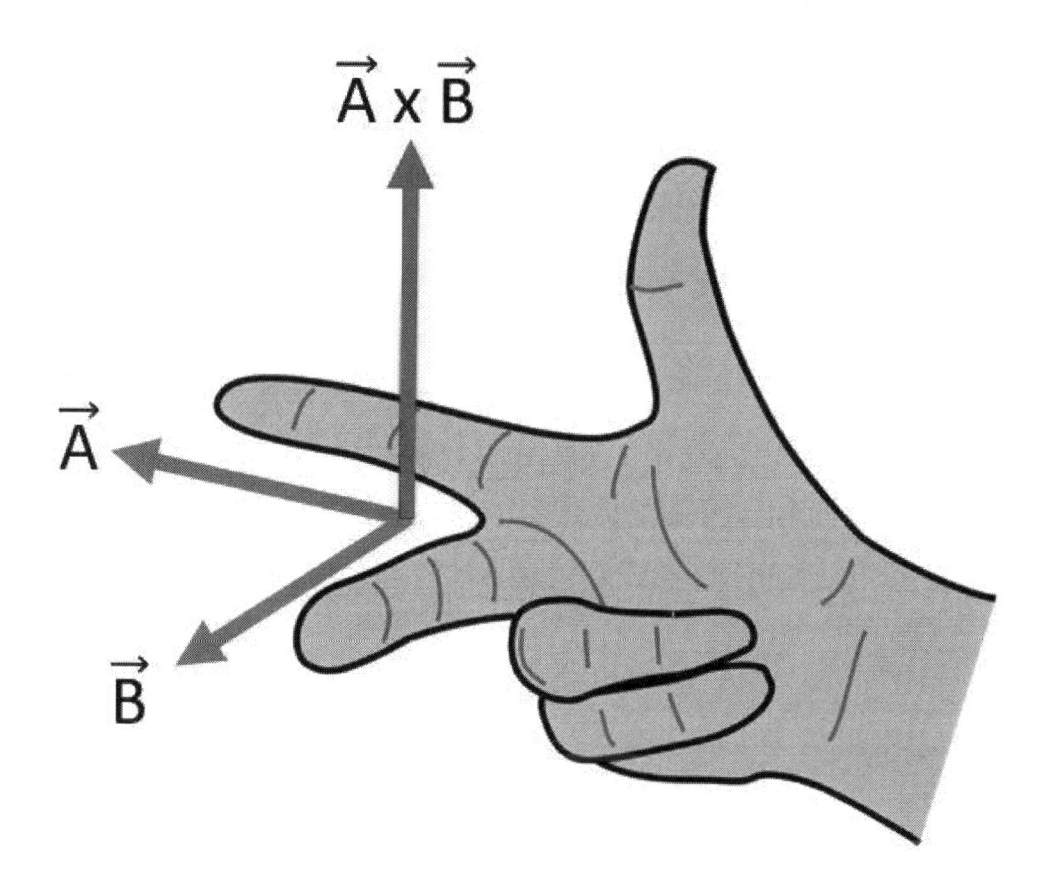

When the angle between two vectors is 0°, the cross product is also zero. And when the angle between the vectors is 90°, the cross product gives a vector with its magnitude equal to the product of their magnitudes. Therefore, the cross products of the 3 fundamental unit vectors are,

$$i \times j = k, \; j \times k = i, \; k \times i = j$$
$$i \times i = j \times j = k \times k = 0$$

Properties of Cross Product:

- A x B = -B x A
- **Distributive property:** A x (B + C) = (A x B) + (A x C)
- k (A x B) = (kA) x B = A x (kB), where **k** is a scalar

- A x A = 0

Solved Examples:

4. Obtain the cross product of vectors A = $\hat{i} + 2\hat{j} - 3\hat{k}$ and B = $3\hat{i} + 5\hat{j} + 7\hat{k}$.

$\vec{A} = \hat{i} + 2\hat{j} - 3\hat{k}, \vec{B} = 3\hat{i} + 5\hat{j} + 7\hat{k}$

$\vec{A} \times \vec{B} = (\hat{i} + 2\hat{j} - 3\hat{k}) \times (3\hat{i} + 5\hat{j} + 7\hat{k})$

$= \begin{vmatrix} \hat{i} & \hat{j} & \hat{k} \\ 1 & 2 & -3 \\ 3 & 5 & 7 \end{vmatrix}$ ← Vectors can be arranged as following & the Determinant gives the Cross product

$= \hat{i}(14+15) - \hat{j}(7+9) + \hat{k}(5-6)$
$= 29\hat{i} - 16\hat{j} - \hat{k}$

5. Obtain the unit vector perpendicular to vectors A = $\hat{i} + 2\hat{j} + \hat{k}$ and B = $3\hat{i} - 4\hat{j} + 2\hat{k}$.

$$\vec{A} = \hat{i} + 2\hat{j} + \hat{k}, \vec{B} = 3\hat{i} - 4\hat{j} + 2\hat{k}$$

$$\vec{A} \times \vec{B} = (\hat{i} + 2\hat{j} + \hat{k}) \times (3\hat{i} - 4\hat{j} + 2\hat{k})$$

$$= \begin{vmatrix} \hat{i} & \hat{j} & \hat{k} \\ 1 & 2 & 1 \\ 3 & -4 & 2 \end{vmatrix}$$

$$= \hat{i}(4+4) - \hat{j}(2-3) + \hat{k}(-4-6)$$

$$= 8\hat{i} + \hat{j} - 10\hat{k}$$

$$\text{Unit vector} = \frac{\vec{A} \times \vec{B}}{|\vec{A} \times \vec{B}|} = \frac{8\hat{i} + \hat{j} - 10\hat{k}}{\sqrt{165}}$$

Note that this unit vector is the upward pointing one, the negative of this vector is also a solution to the above problem.

6. A torque is applied on a nut using a wrench as shown in Figure. Find the magnitude of torque applied to the nut and find whether the nut is being tightened or loosened? Also find the maximum torque that can be applied to the nut with the same magnitude of force.

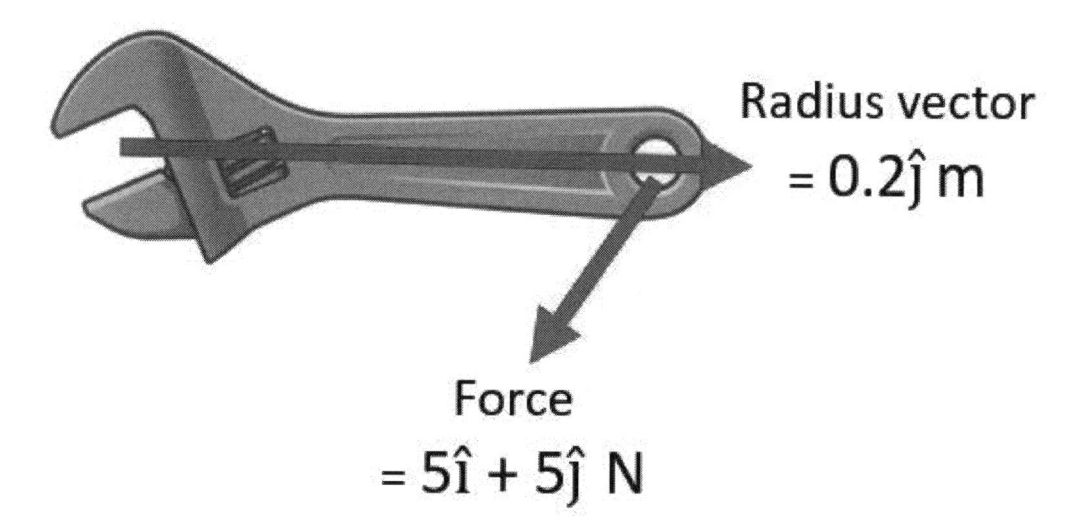

Radius vector $= 0.2\hat{j}$ m

Force $= 5\hat{i} + 5\hat{j}$ N

Torque is the twisting action or moment of a force. It is equal to the cross product of the lever arm vector and the Force applied.

$$\text{Torque } \vec{\tau} = \vec{r} \times \vec{F}$$

$$= \begin{vmatrix} \hat{i} & \hat{j} & \hat{k} \\ 0 & 0.2 & 0 \\ 5 & 5 & 0 \end{vmatrix}$$

$$= -\hat{k}$$

$$|\vec{\tau}| = 1 \text{ Nm, Downwards}$$

As the Torque is in the downward direction, the nut is being tightened using the wrench.

$$|\vec{F}| = \sqrt{5^2 + 5^2} = 5\sqrt{2}$$

$$\tau_{max} = |\vec{r} \times \vec{F}|_{\theta = 90°}$$
$$= |\vec{r}||\vec{F}|$$
$$= (0.2)(5\sqrt{2}) = 1.41 \text{ Nm}$$

2.3 SCALAR TRIPLE PRODUCT

So far we dealt with the product of 2 vectors, now we move on to product of 3 vectors, called the triple product. First of these is the Scalar triple product. It is the dot product of 2 vectors, one of which itself is the cross product of 2 other vectors.

$$\vec{A} \cdot (\vec{B} \times \vec{C})$$

The most meaningful way to interpret the scalar triple product is that it represents the volume of a parallelopiped formed by 3 vectors taken as its adjacent sides.

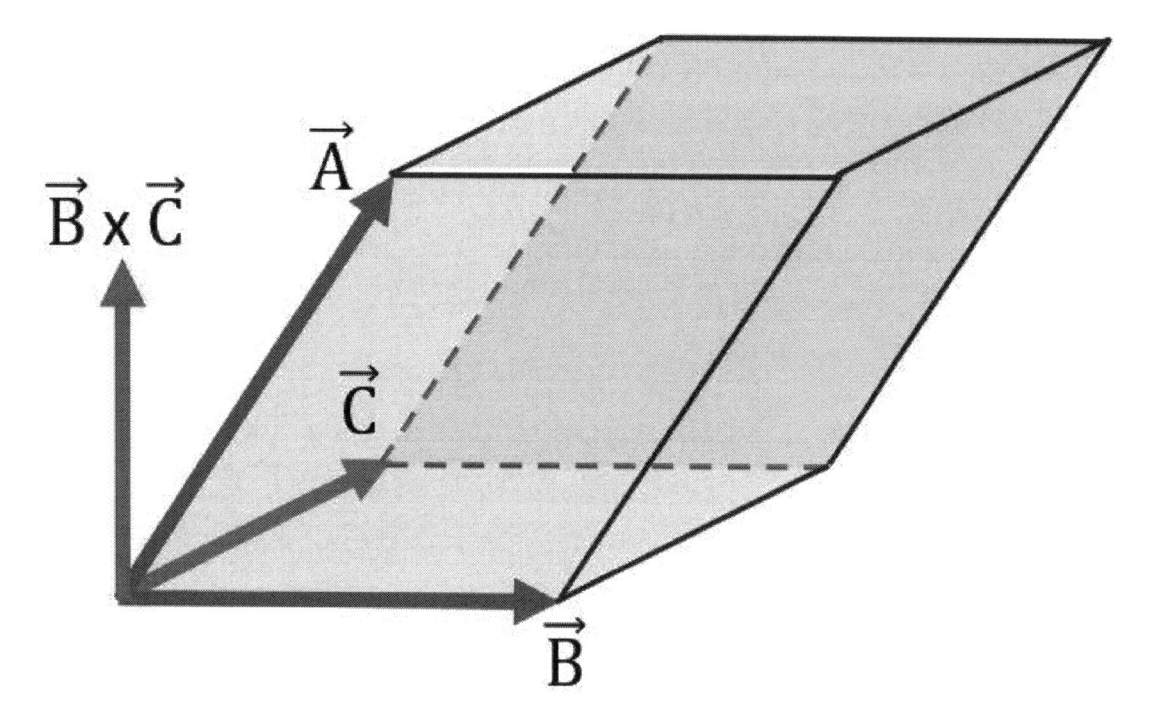

Here the vectors B and C form the base of the parallelopiped. The cross product B x C hence represents the area vector of the base. Vector A denotes the slant height of the parallelopiped, therefore A.(BxC), which is the projection of the area vector on the slant height, represents the volume of the parallelopiped. Now what if the vector A was downward facing? In that case the volume would have been negative. Do not read too much into this, it simply means that if 3 vectors form a left-handed system, then the corresponding scalar triple product will give a negative value.

The best part about this geometrical interpretation is that several properties of the scalar triple product can be deduced from it without much messy work. For instance, if all 3 vectors A, B and C were coplanar, then no parallelopiped can be formed using them and as result the scalar triple product would be zero. In fact this property is used to check the coplanarity of 3 vectors.

Properties of Scalar Triple Product:

- **Commutative property: A.(B x C) = (B x C).A**

Since the dot product is commutative, the scalar triple product too is commutative in nature.

- **A.(B x C) = B.(C x A) = C.(A x B)**

This property can be proven using the fact that irrespective of which 2 vectors use choose as the base for the parallelopiped, its volume will remain the same. The only thing to be kept in mind is that the cyclic order (A-B-C-A) must be maintained.

From this property and the commutative property, we can conclude that the Scalar triple product has 6 different forms.

$$A \cdot (B \times C) = B \cdot (C \times A) = C \cdot (A \times B)$$

$$= (B \times C) \cdot A = (C \times A) \cdot B = (A \times B) \cdot C$$

All of these forms can be conveniently denoted using the symbol **[A B C]**.

- The scalar triple product of the 3 fundamental unit vectors is,

$$[\hat{\imath} \ \hat{\jmath} \ \hat{k}] = \hat{\imath} \cdot (\hat{\jmath} \times \hat{k})$$
$$= \hat{\imath} \cdot \hat{\imath} = 1$$

If 3 vectors are expressed in terms of the fundamental unit vectors, then their scalar triple product can be expressed as a determinant as follows,

$$\vec{A} = a_1 \hat{\imath} + b_1 \hat{\jmath} + c_1 \hat{k}$$
$$\vec{B} = a_2 \hat{\imath} + b_2 \hat{\jmath} + c_2 \hat{k}$$
$$\vec{C} = a_3 \hat{\imath} + b_3 \hat{\jmath} + c_3 \hat{k}$$

$$[\vec{A} \ \vec{B} \ \vec{C}] = \begin{vmatrix} a_1 & b_1 & c_1 \\ a_2 & b_2 & c_2 \\ a_3 & b_3 & c_3 \end{vmatrix}$$

Solved Examples:

7. Prove that the vectors $A = \hat{i} + 2\hat{j}$, $B = 3\hat{i} + 5\hat{j}$ and $C = 10\hat{i}$ are coplanar.

For 3 coplanar vectors, their scalar triple product will be zero.

$$[\vec{A}\ \vec{B}\ \vec{C}] = \begin{vmatrix} 1 & 2 & 0 \\ 3 & 5 & 0 \\ 10 & 0 & 0 \end{vmatrix}$$

$$= 1(0-0) - 2(0-0) + 0(0-50)$$

$$= 0$$

8. Find the volume of the parallelopiped spanned by the vectors $A = (-2,3,1)$, $B = (0,4,0)$, and $C = (-1,3,3)$.

The scalar triple product represents the volume of a parallelopiped spanned by 3 vectors and its absolute value gives the actual volume.

$$[\vec{A}\ \vec{B}\ \vec{C}] = \begin{vmatrix} -1 & 3 & 3 \\ -2 & 3 & 1 \\ 0 & 4 & 0 \end{vmatrix}$$

$$= -1(0-4) - 3(0-0) + 3(8-0)$$

$$= -20$$

Hence the Volume of the parallelopiped = $|-20|$ = 20 Units.

9. If the vectors $A = \hat{i} + \hat{j} + \hat{k}$, $B = \hat{i}$ and $C = \hat{i} + 2\hat{j} + a\hat{k}$ are coplanar. Find the value of a?

For 3 coplanar vectors, their scalar triple product will be zero.

$$[\vec{A}\ \vec{B}\ \vec{C}] = \begin{vmatrix} 1 & 1 & 1 \\ 1 & 0 & 0 \\ 1 & 2 & a \end{vmatrix} = 0$$

$$= 1(0-0) - 1(a-0) + 1(2-0)$$

$$= -a + 2 = 0$$

$$\therefore a = 2$$

2.4 VECTOR TRIPLE PRODUCT

The second type of triple product is the vector triple product. It is the cross of two vectors, one of which is itself is the cross product of 2 other vectors.

$$\vec{A} \times (\vec{B} \times \vec{C})$$

Unfortunately, there isn't an easy way to interpret the vector triple product. But there is a way to visualize everything that's going on.

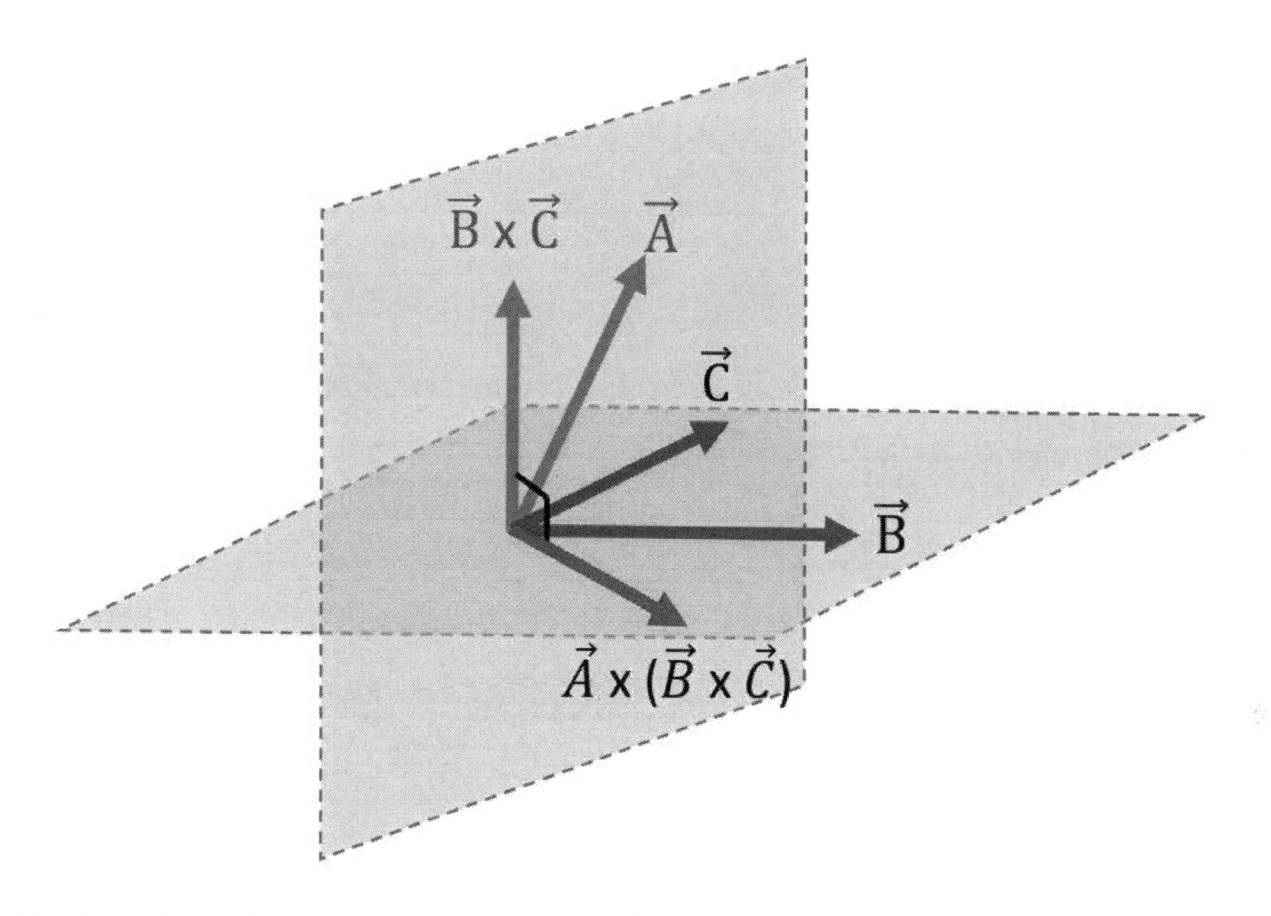

We know that the cross product vector of 2 vectors is perpendicular to both vectors, therefore the vector B x C is perpendicular to the plane of B and C. Moreover, the vector A x (B x C) is perpendicular to both vector A and the vector B x C. This means that it must lie in the plane of B and C as shown in the figure above.

Properties of the Vector Triple Product:

- **A x (B x C) ≠ (A x B) x C**

Vector Triple product is not Associative. The reasoning is simple, A x (B x C) vector will be coplanar with vectors B and C, whereas (A x B) x C vector will be coplanar with vectors A and B. So clearly both these vectors cannot be the same.

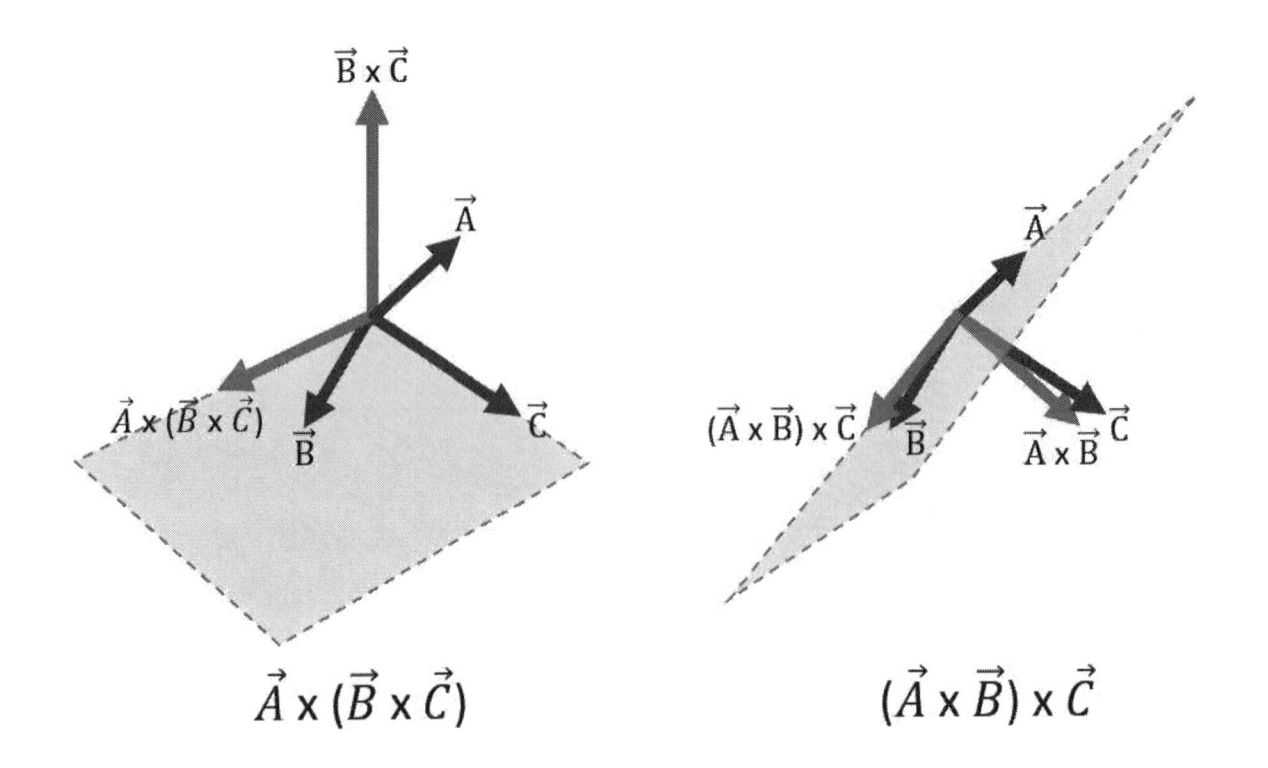

$$\vec{A} \times (\vec{B} \times \vec{C}) \qquad\qquad (\vec{A} \times \vec{B}) \times \vec{C}$$

- **A x (B x C) = (A.C)B − (A.B)C**

The vector triple product can be expressed as the difference of 2 dot products as given. The easy way to remember this expansion is by using the mnemonic "ACB − ABC", also keeping in mind that the first 2 vectors are dotted together. The proof for this result is provided in the appendix.

Solved Examples:

10. Find the Vector triple product A x (B x C) of the vectors A = (−2,3,1), B = (0,4,0), and C = (−1,3,3). Also find (A x B) x C and show that they are not equal.

$$\vec{B} \times \vec{C} = \begin{vmatrix} \hat{\imath} & \hat{\jmath} & \hat{k} \\ 0 & 4 & 0 \\ -1 & 3 & 3 \end{vmatrix}$$

$$= 12\hat{\imath} + 4\hat{k}$$

$$\vec{A} \times (\vec{B} \times \vec{C}) = \begin{vmatrix} \hat{\imath} & \hat{\jmath} & \hat{k} \\ -2 & 3 & 1 \\ 12 & 0 & 4 \end{vmatrix}$$

$$= 12\hat{\imath} + 20\hat{\jmath} - 36\hat{k}$$

$$\vec{A} \times \vec{B} = \begin{vmatrix} \hat{\imath} & \hat{\jmath} & \hat{k} \\ -2 & 3 & 1 \\ 0 & 4 & 0 \end{vmatrix}$$

$$= -4\hat{\imath} - 8\hat{k}$$

$$(\vec{A} \times \vec{B}) \times \vec{C} = \begin{vmatrix} \hat{\imath} & \hat{\jmath} & \hat{k} \\ -4 & 0 & -8 \\ -1 & 3 & 3 \end{vmatrix}$$

$$= 24\hat{\imath} + 20\hat{\jmath} - 12\hat{k}$$

Therefore **A x (B x C) ≠ (A x B) x C**

11. **Prove that A x (B x C) = (A.C) B – (A.B) C, using vectors A = (1,1,1), B = (2,1,0), and C = (4,1,2).**

$$\vec{B} \times \vec{C} = \begin{vmatrix} \hat{\imath} & \hat{\jmath} & \hat{k} \\ 2 & 1 & 0 \\ 4 & 1 & 2 \end{vmatrix}$$

$$= 2\hat{\imath} - 4\hat{\jmath} - 2\hat{k}$$

$$\vec{A} \times (\vec{B} \times \vec{C}) = \begin{vmatrix} \hat{\imath} & \hat{\jmath} & \hat{k} \\ 1 & 1 & 1 \\ 2 & -4 & -2 \end{vmatrix}$$

LHS $= 2\hat{\imath} + 4\hat{\jmath} - 6\hat{k}$

$\vec{A}.\vec{C} = 7$
$(\vec{A}.\vec{C})\vec{B} = 14\hat{\imath} + 7\hat{\jmath}$

$\vec{A}.\vec{B} = 3$
$(\vec{A}.\vec{B})\vec{C} = 12\hat{\imath} + 3\hat{\jmath} + 6\hat{k}$

$\therefore (\vec{A}.\vec{C})\vec{B} - (\vec{A}.\vec{B})\vec{C} = 2\hat{\imath} + 4\hat{\jmath} - 6\hat{k}$
$\phantom{\therefore (\vec{A}.\vec{C})\vec{B} - (\vec{A}.\vec{B})\vec{C} \;\;} =$ RHS

Clearly LHS = RHS and hence the identity **A x (B x C) = (A.C) B – (A.B) C** stands.

2.5 RECIPROCAL VECTORS

For any 3 non coplanar vectors **a**, **b** and **c**, another set of 3 vectors **a'**, **b'** and **c'** called the Reciprocal system of vectors can be defined. They can be obtained as,

$$\vec{a'} = \frac{\vec{b} \times \vec{c}}{[\vec{a}\ \vec{b}\ \vec{c}]}, \quad \vec{b'} = \frac{\vec{c} \times \vec{a}}{[\vec{a}\ \vec{b}\ \vec{c}]}, \quad \vec{c'} = \frac{\vec{a} \times \vec{b}}{[\vec{a}\ \vec{b}\ \vec{c}]}$$

The main advantage of reciprocal vectors is that it allows us to reconstruct a vector from its dot products with any 3 non coplanar vectors,

$$\vec{A} = (\vec{A} \cdot \vec{a})\ \vec{a'} + (\vec{A} \cdot \vec{b})\ \vec{b'} + (\vec{A} \cdot \vec{c})\ \vec{c'}$$

Properties of Reciprocal Vectors:

- The dot product of any vector with its corresponding reciprocal vector is unity and the dot product of any vector with its non-corresponding reciprocal vector is zero.

$$\vec{a'} \cdot \vec{a} = \vec{b'} \cdot \vec{b} = \vec{c'} \cdot \vec{c} = 1$$

$$\vec{a'} \cdot \vec{b} = \vec{a'} \cdot \vec{c} = \vec{b'} \cdot \vec{a} = \vec{b'} \cdot \vec{c} = \vec{a'} \cdot \vec{b} = \vec{c'} \cdot \vec{a} = \vec{c'} \cdot \vec{b} = 0$$

- The 3 fundamental unit vectors **i, j, k** is its own reciprocal system.

$$\hat{i'} = \frac{\hat{j} \times \hat{k}}{[\hat{i}\ \hat{j}\ \hat{k}]} = \frac{\hat{i}}{1} = \hat{i}$$

$$\hat{j'} = \frac{\hat{k} \times \hat{i}}{[\hat{i}\ \hat{j}\ \hat{k}]} = \frac{\hat{j}}{1} = \hat{j}$$

$$\hat{k'} = \frac{\hat{i} \times \hat{j}}{[\hat{i}\ \hat{j}\ \hat{k}]} = \frac{\hat{k}}{1} = \hat{k}$$

Solved Examples:

12. Find the reciprocal set of vectors to vectors a = <2,3,-1>, b = <1,-1,-2> and c = <-1,2,2>.

$$\vec{a} \times \vec{b} = \begin{vmatrix} \hat{\imath} & \hat{\jmath} & \hat{k} \\ 2 & 3 & -1 \\ 1 & -1 & -2 \end{vmatrix} = -7\hat{\imath} + 3\hat{\jmath} - 5\hat{k}$$

$$\vec{b} \times \vec{c} = \begin{vmatrix} \hat{\imath} & \hat{\jmath} & \hat{k} \\ 1 & -1 & -2 \\ -1 & 2 & 2 \end{vmatrix} = 2\hat{\imath} + \hat{k}$$

$$\vec{c} \times \vec{a} = \begin{vmatrix} \hat{\imath} & \hat{\jmath} & \hat{k} \\ -1 & 2 & 2 \\ 2 & 3 & -1 \end{vmatrix} = -8\hat{\imath} + 3\hat{\jmath} - 7\hat{k}$$

$$[\vec{a}\ \vec{b}\ \vec{c}] = \begin{vmatrix} 2 & 3 & -1 \\ 1 & -1 & -2 \\ -1 & 2 & 2 \end{vmatrix} = 3$$

$$\therefore \vec{a}' = \frac{2\hat{\imath} + \hat{k}}{3},\ \vec{b}' = \frac{-8\hat{\imath} + 3\hat{\jmath} - 7\hat{k}}{3},\ \vec{c}' = \frac{-7\hat{\imath} + 3\hat{\jmath} - 5\hat{k}}{3}$$

13. Find the reciprocal set of vectors to vectors a = <1,1,0>, b = <0,1,0> and c = <1,1,1>.

40

$$\vec{a} \times \vec{b} = \begin{vmatrix} \hat{i} & \hat{j} & \hat{k} \\ 1 & 1 & 0 \\ 0 & 1 & 0 \end{vmatrix} = \hat{k}$$

$$\vec{b} \times \vec{c} = \begin{vmatrix} \hat{i} & \hat{j} & \hat{k} \\ 0 & 1 & 0 \\ 1 & 1 & 1 \end{vmatrix} = \hat{i} - \hat{k}$$

$$\vec{c} \times \vec{a} = \begin{vmatrix} \hat{i} & \hat{j} & \hat{k} \\ 1 & 1 & 1 \\ 1 & 1 & 0 \end{vmatrix} = -\hat{i} + \hat{j}$$

$$[\vec{a} \ \vec{b} \ \vec{c}] = \begin{vmatrix} 1 & 1 & 0 \\ 0 & 1 & 0 \\ 1 & 1 & 1 \end{vmatrix} = 1$$

$$\therefore \vec{a'} = \frac{\vec{b} \times \vec{c}}{[\vec{a} \ \vec{b} \ \vec{c}]} = \hat{i} - \hat{k}, \quad \vec{b'} = \frac{\vec{c} \times \vec{a}}{[\vec{a} \ \vec{b} \ \vec{c}]} = -\hat{i} + \hat{j},$$

$$\vec{c'} = \frac{\vec{a} \times \vec{b}}{[\vec{a} \ \vec{b} \ \vec{c}]} = \hat{k}$$

3. VECTOR DIFFERENTIATION

From this chapter onwards we are dealing with Vector calculus, so in the next few sections we'll review some related mathematical concepts.

3.1 LIMIT

The limit of a function describes the behavior of a function in the neighborhood of a point, but not the point itself. Consider a function **f(x)** as shown below.

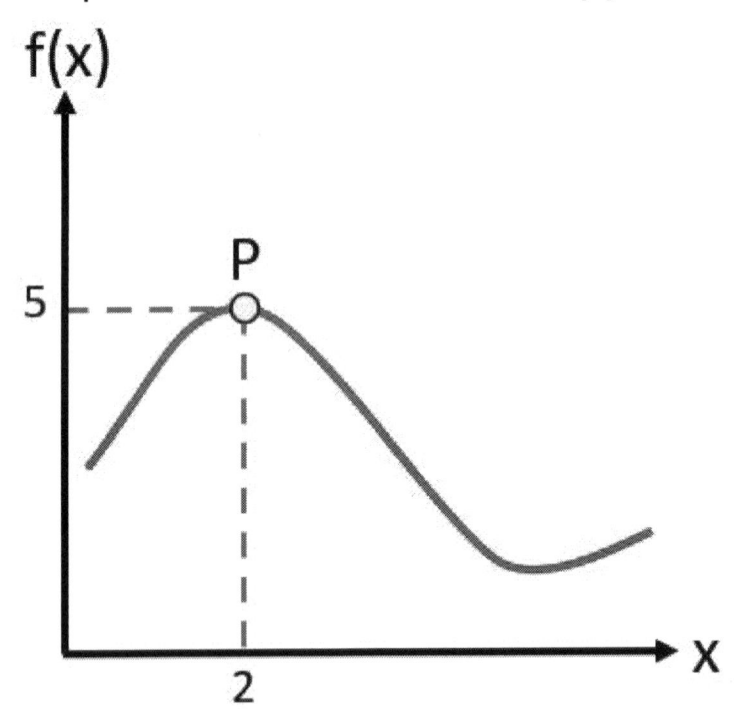

At **x =2**, the value of the function is 5, denoted by point P on the curve. Suppose we want to obtain the value of the function at a point in the close proximity of point P, say at **x = 1.99** or at **x = 2.01**, for that we use limits. The limit of a function is represented as,

$$\lim_{x \to x_0} f(x)$$

It's read as "Limit **f(x)** as **x** tends to **x₀**".

So how close should a point be to be considered as "in the neighborhood" of another point? Technically there isn't such a defined neighborhood, but depending on the range of the function it can be chosen appropriately. For a single variable function as in our example, a point can be approached in 2 directions, from the left and from the right. The limit of a function as the point is approached from the left side is called the Left hand limit and similarly, when approached from the right side it's called the Right hand limit.

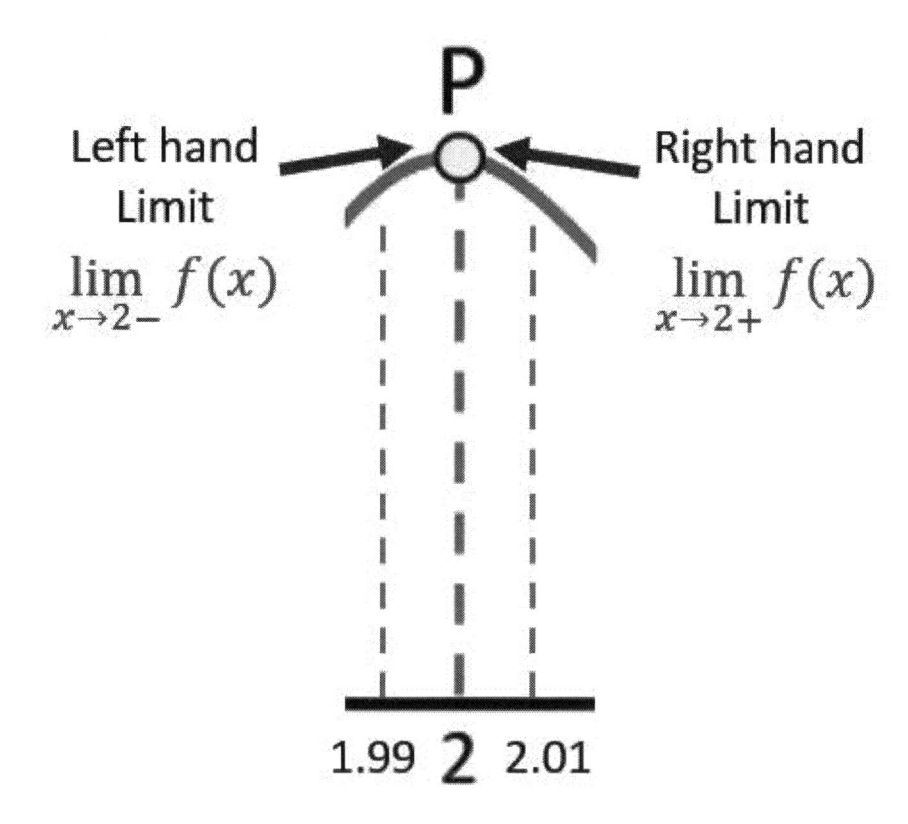

Left hand Limit
$\lim_{x \to 2-} f(x)$

P

Right hand Limit
$\lim_{x \to 2+} f(x)$

1.99 2 2.01

The important thing to note here is that the value of the function as it approaches a point need not be the same as the value of the function at the point. In fact it is possible that the function may not even be defined at a point and still have values in its neighborhood. In other cases, the function may have different values when approached from the left and the right side. This is where the concept of continuity comes into play.

A function is said to be continuous if the value of the function at a point is the same as the limit of the function in the neighborhood of that point i.e.

$$\lim_{x \to x_0} f(x) = f(x_0)$$

What this means is that there is no break or a missing point for function **f(x)** at the point **x₀**. In other words, you could draw the graph of **f(x)** through the point **x₀** without lifting the pencil off the paper.

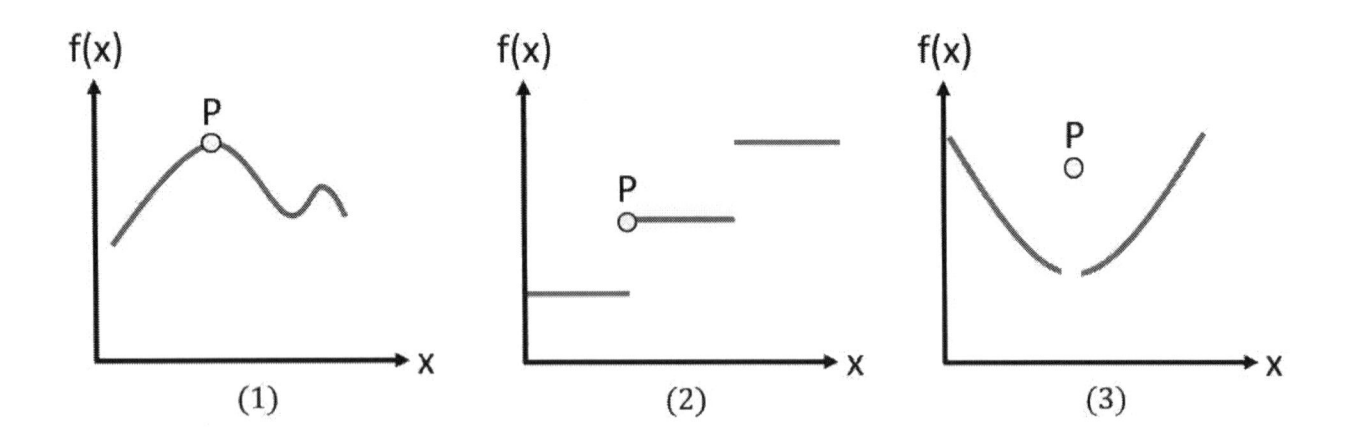

Consider the 3 graphs shown above. In the first graph, the value of the limit is the same as the value of the function at point P and therefore it is continuous at point P. In the second case, the value of the limit as the point P is approached from the left side is not the same as the value of the limit as it is approached from the right side, therefore the limits doesn't exist. For that reason, as obvious from observation, this function is not continuous. The third case is a bit more interesting. Here the value of the limit as the function approaches P from both the right and the left side are the same, so the limit exists. But the value of the limit is not the same as the value of the function at P, so this function too is not continuous. Continuity of a function is defined at a point or a range.

The concept of Limits and Continuity is very important as far as calculus is concerned.

3.2 DERIVATIVE

In calculus, we are dealing with quantities that change and by how much they are changing etc. Consider a function **f(x)** as shown below.

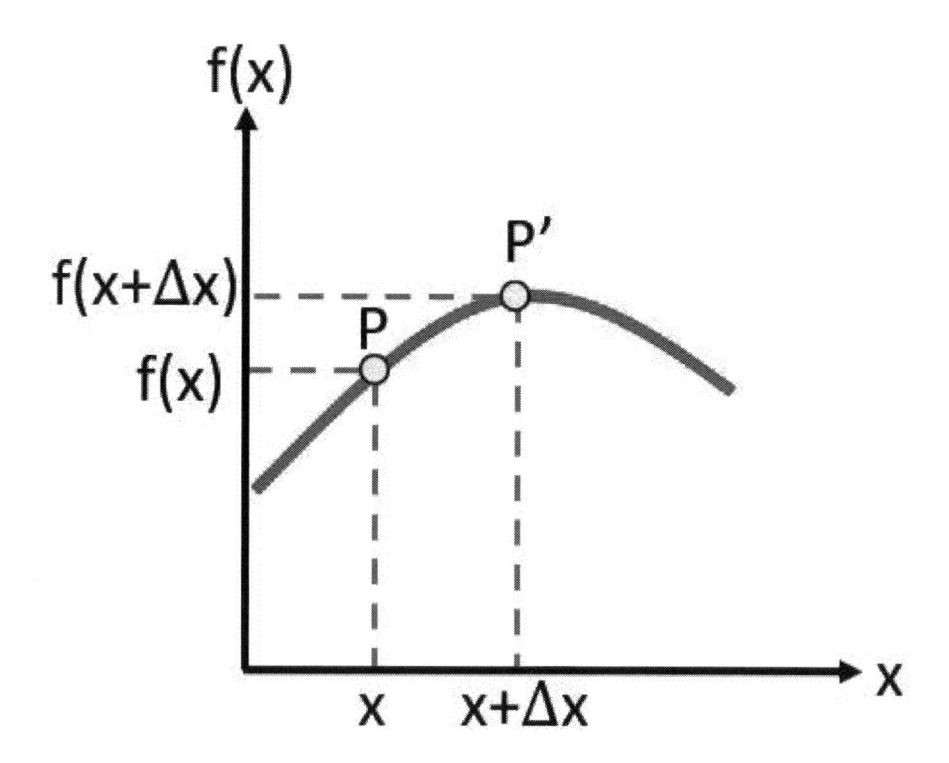

In this function, as **x** changes from **x** to **x+Δx**, the corresponding **f(x)** value changes from **f(x)** to **f(x+Δx)**, as denoted by the points P and P' respectively. So the change in **f(x)** with respect to change in **x** can be expressed as,

$$\frac{f(x+\Delta x)-f(x)}{(x+\Delta x)-(x)}$$

Now imagine if we choose **x+Δx** very close to **x** i.e. we limit **Δx** to zero, then as a result the points P and P' would almost merge. Almost! By doing so what we get is the instantaneous change in the function **f(x)** with respect to **x** at a particular point (In this case point P). That is the derivative. It is denoted as **f'(x)** (read as **f prime x**).

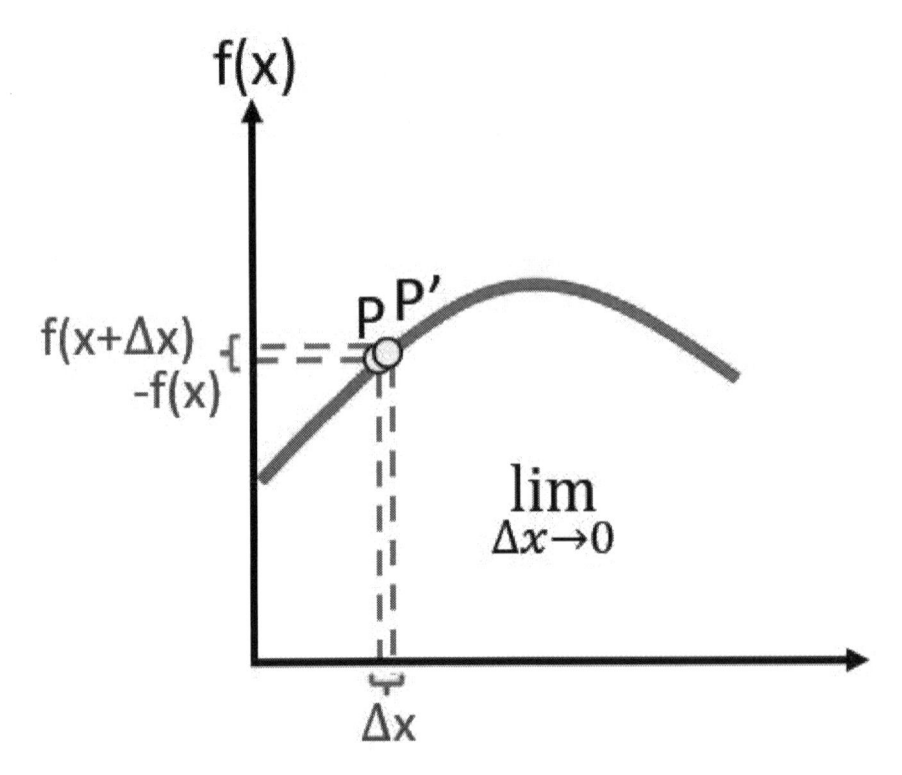

Therefore, the derivative of **f(x)** with respect **x** can be expressed as,

$$f'(x) = \lim_{\Delta x \to 0} \frac{f(x+\Delta x) - f(x)}{\Delta x}$$

Geometrically, the derivative denotes the Slope of the tangent at a point on the curve. More the slope, faster the function **f(x)** increases at that point.

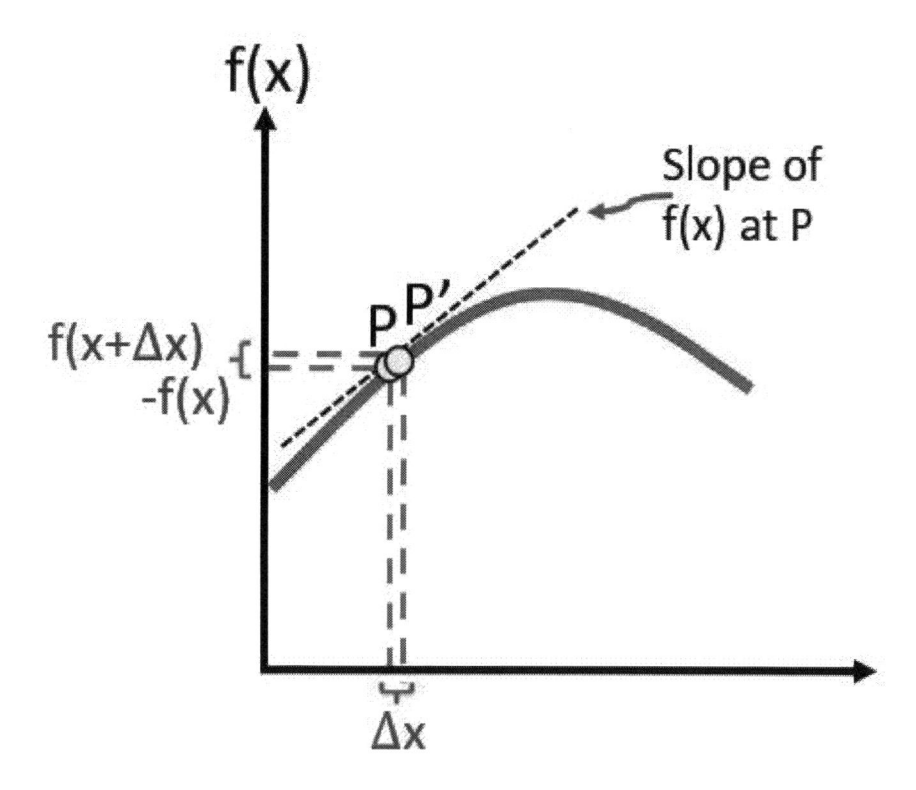

Since our focus is on vectors, we'll leave it at this.

3.3 DERIVATIVE OF A VECTOR

In the previous 2 chapters we dealt with vectors that were constant. But in the physical world quantities are often varying and so are the vectors describing them. In most cases, Vectors are usually a function of time like in the case of displacement or a function of position like with Electric field intensity. Generally, a vector which is a function of a scalar quantity (**u**) can be expressed as **V= f(u)**.

So what does the derivative mean in the context of vectors? Consider the motion of a particle as shown below.

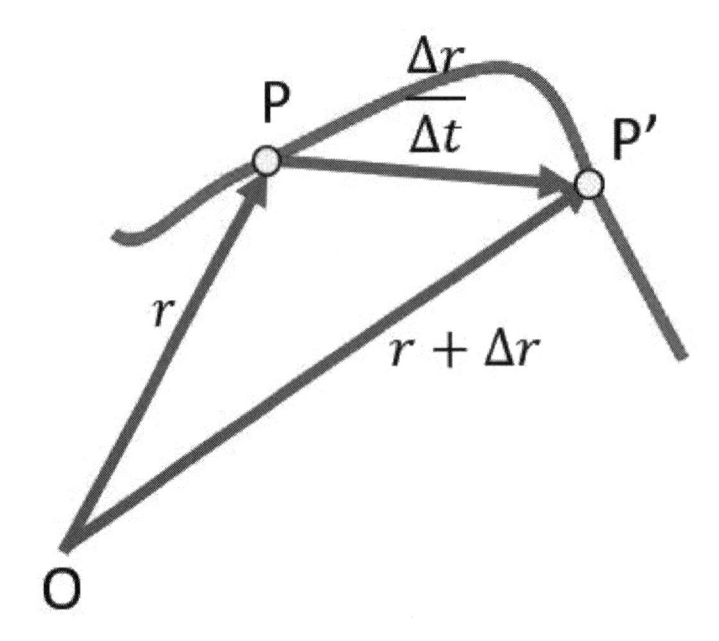

At time **t**, the particle is at point P denoted by the vector **r** (with reference to some point, let's say the origin), then at another instance of time **t+Δt**, the particle has now moved to position P' denoted by the vector **r+Δr**. Therefore, the vector PP' denotes the change in position with respect to change in time.

$$\frac{Change\ in\ Position}{Change\ in\ time} = \frac{(r+\Delta r)-r}{(t+\Delta t)-t} = \frac{\Delta r}{\Delta t}$$

Now if we chose a very small interval of time (**Δt→0**), say a split second apart, obviously the particle wouldn't have moved much and therefore the point P' would almost converge to point P (**Δr→0**). And what we get as a result is the derivative of the vector **r** with respect to time **t** at point P. To be more specific, it denotes the instantaneous rate of change of the vector with respect to time.

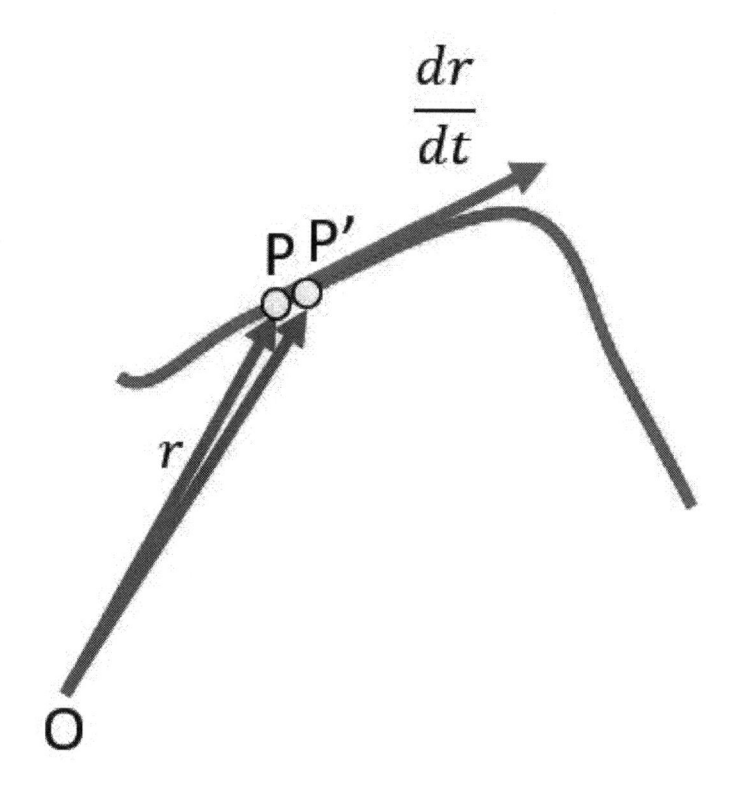

$$\frac{d\vec{r}}{dt} = \lim_{\Delta t \to 0} \frac{\Delta \vec{r}}{\Delta t}$$

If we expressed the vector **r** in terms of unit vectors **i, j** and **k** as,

$$\vec{r} = r_x \hat{i} + r_y \hat{j} + r_z \hat{k}$$

then each of these components (r_x, r_y, r_z) will also be functions of the scalar quantity **t**. Therefore,

$$\vec{r} + \Delta\vec{r} = (r_x + \Delta r_x)\hat{i} + (r_y + \Delta r_y)\hat{j} + (r_z + \Delta r_z)\hat{k}$$

$$\Delta\vec{r} = \Delta r_x \hat{i} + \Delta r_y \hat{j} + \Delta r_z \hat{k}$$

$$\frac{\Delta\vec{r}}{\Delta t} = \frac{\Delta r_x}{\Delta t}\hat{i} + \frac{\Delta r_y}{\Delta t}\hat{j} + \frac{\Delta r_z}{\Delta t}\hat{k}$$

$$\boxed{\frac{d\vec{r}}{dt} = \lim_{\Delta t \to 0} \frac{\Delta\vec{r}}{\Delta t} = \frac{d\vec{r}_x}{dt}\hat{i} + \frac{d\vec{r}_y}{dt}\hat{j} + \frac{d\vec{r}_x}{dt}\hat{k}}$$

Hence the components of the derivative of **r** with respect to **t** will be equal to the derivatives of the components of **r** with respect to **t**. Read that again!

This can be extended to higher order derivatives as well.

Solved Examples:

1. **Evaluate the derivative of the vector function r(t)=*ln t* $\hat{i}$ +(*3t+1*) $\hat{j}$ +*t²* $\hat{k}$ at t=1.**

50

$$\vec{r} = \ln t \, \hat{\imath} + (3t+1)\hat{\jmath} + t^2 \hat{k}$$

$$\frac{d\vec{r}}{dt} = \frac{d(\ln t)}{dt}\hat{\imath} + \frac{d(3t+1)}{dt}\hat{\jmath} + \frac{d(t^2)}{dt}\hat{k}$$

$$= \frac{1}{t}\hat{\imath} + 3\hat{\jmath} + 2t\,\hat{k}$$

At t =1,
$$\vec{r}'(t) = \hat{\imath} + 3\hat{\jmath} + 2\hat{k}$$

2. A particle moves along the curve r(t)= ⟨ t^2, t^2−4t, t ⟩. Find the speed of the particle at t=3.

$$\vec{r} = t^2\hat{\imath} + (t^2 - 4t)\hat{\jmath} + t\,\hat{k}$$

$$\frac{d\vec{r}}{dt} = 2t\,\hat{\imath} + (2t-4)\hat{\jmath} + \hat{k}$$

At t =3,
$$\vec{v}(3) = 6\hat{\imath} + 2\hat{\jmath} + \hat{k}$$

$$|\vec{v}(1)| = \sqrt{6^2 + 2^2 + 1^2} = 6.4$$

3.4 DERIVATIVE OF SUM & PRODUCTS OF VECTORS

Here are some properties of vector differentiation.

1.
$$(A \pm B)' = A' \pm B'$$

Proof:

Let **A** and **B** be 2 vectors that are functions of a scalar quantity **t**. Then, if **t** changes by an amount Δt, let vector **A** and **B** change by Δ**A** & Δ**B** respectively.

$$\frac{\Delta(A + B)}{\Delta t} = \frac{((A + \Delta A) + (B + \Delta B)) - (A + B)}{\Delta t}$$

$$= \frac{\Delta A + \Delta B}{\Delta t}$$

$$\frac{d(A + B)}{dt} = \lim_{\Delta t \to 0} \frac{\Delta A + \Delta B}{\Delta t}$$

$$= \frac{dA}{dt} + \frac{dB}{dt}$$

2.   (A.B)' = A'.B + A'.B

Proof:

$$\frac{\Delta(A.B)}{\Delta t} = \frac{((A + \Delta A).(B + \Delta B)) - (A.B)}{\Delta t}$$

$$= \frac{(A.\Delta B + \Delta A.B + \Delta A.\Delta B)}{\Delta t}$$

$$= \frac{A.\Delta B}{\Delta t} + \frac{\Delta A.B}{\Delta t} + \frac{\Delta A.\Delta B}{\Delta t}$$

$$\frac{d(A.B)}{dt} = \lim_{\Delta t \to 0} \frac{A.\Delta B}{\Delta t} + \frac{\Delta A.B}{\Delta t} + \frac{\Delta A.\Delta B}{\Delta t}$$

$$= A\frac{dB}{dt} + B\frac{dA}{dt}$$

The last term in the expansion can be neglected as both **ΔA** & **ΔB** are very small quantities and hence their product (**ΔA.ΔB**) is an even smaller quantity.

3. (A x B)' = A' x B + A' x B

The proof of this property is identical to the above one. But one thing that should be kept in your mind is that wherever the cross product is involved the original order of the factors must be maintained throughout the equation. That is, (A x B)' ≠ A' x B + B' x A. The reason being that the cross product is not commutative in nature.

3.5 SCALAR AND VECTOR FIELDS

We saw vectors that were functions of a single scalar quantity, now we'll turn our attention to vectors that are functions of multiple scalar quantities. We are interested in "fields" in particular. So what is a field?

A field is a physical quantity that can be specified everywhere in space as a function of position (**x**, **y** and **z** coordinates). And there are basically 2 types of fields: Scalar and vector fields.

A scalar field is a function that associates a single scalar value or a magnitude to every point in space. For example, the Distribution of temperature in a room. If Temperature in a room is given by the scalar field **T= xy^2z^3**, then at point (1,1,1) the temperature is 1 unit and at another point (2,3,1), the temperature is 18 units and so on. (Shown below is the predicted heatmap of the globe in 2100, this is also an example of a scalar field)

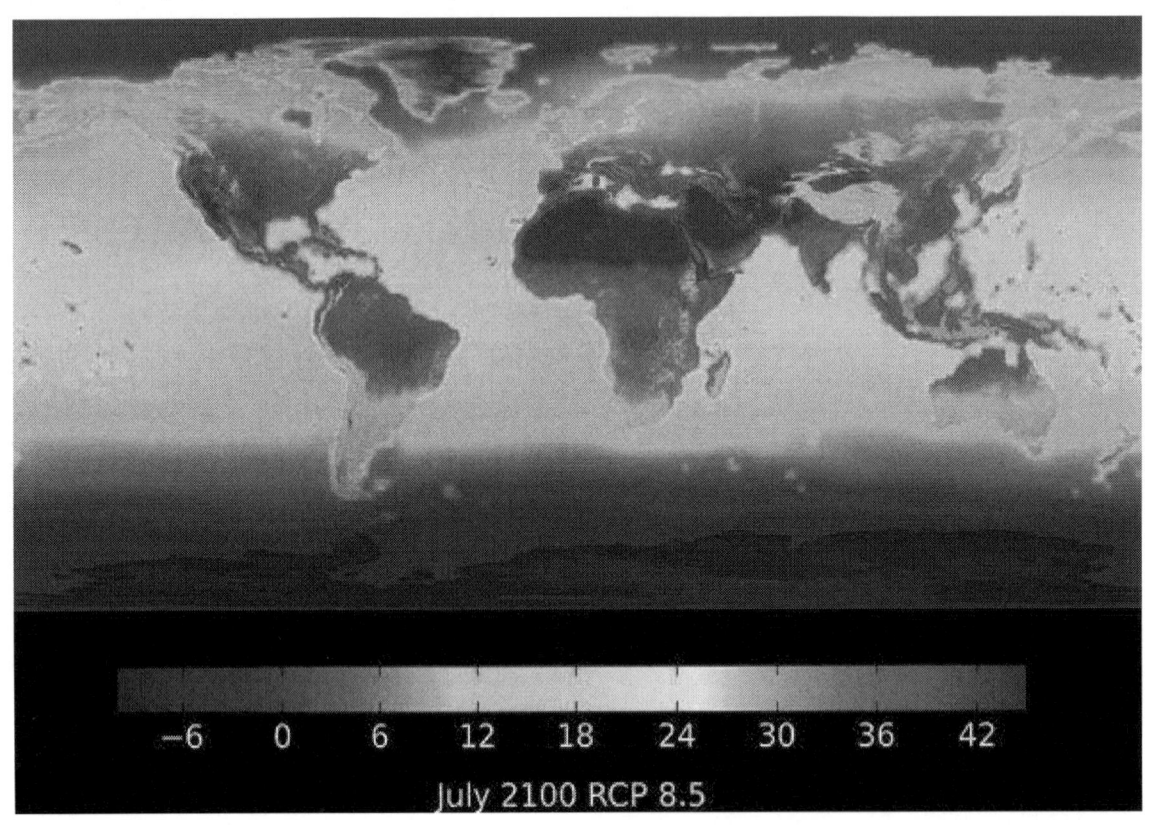

Note that it is not necessary for a scalar field to have non zero values at every point in space.

Similarly, a vector field is a function that associates a vector value to every point in space. For example, the velocity of flow at different points in a fluid. If velocity of flow is given by **V = $xy\,\hat{i} + xy^2z\,\hat{j} - z^3\,\hat{k}$**, then at point (1,1,1), the velocity is denoted by the vector $\hat{i} + \hat{j} - \hat{k}$. At another point (1,2,3), the velocity is denoted by vector **$2\hat{i} + 12\hat{j} - 27\hat{k}$** and so on.

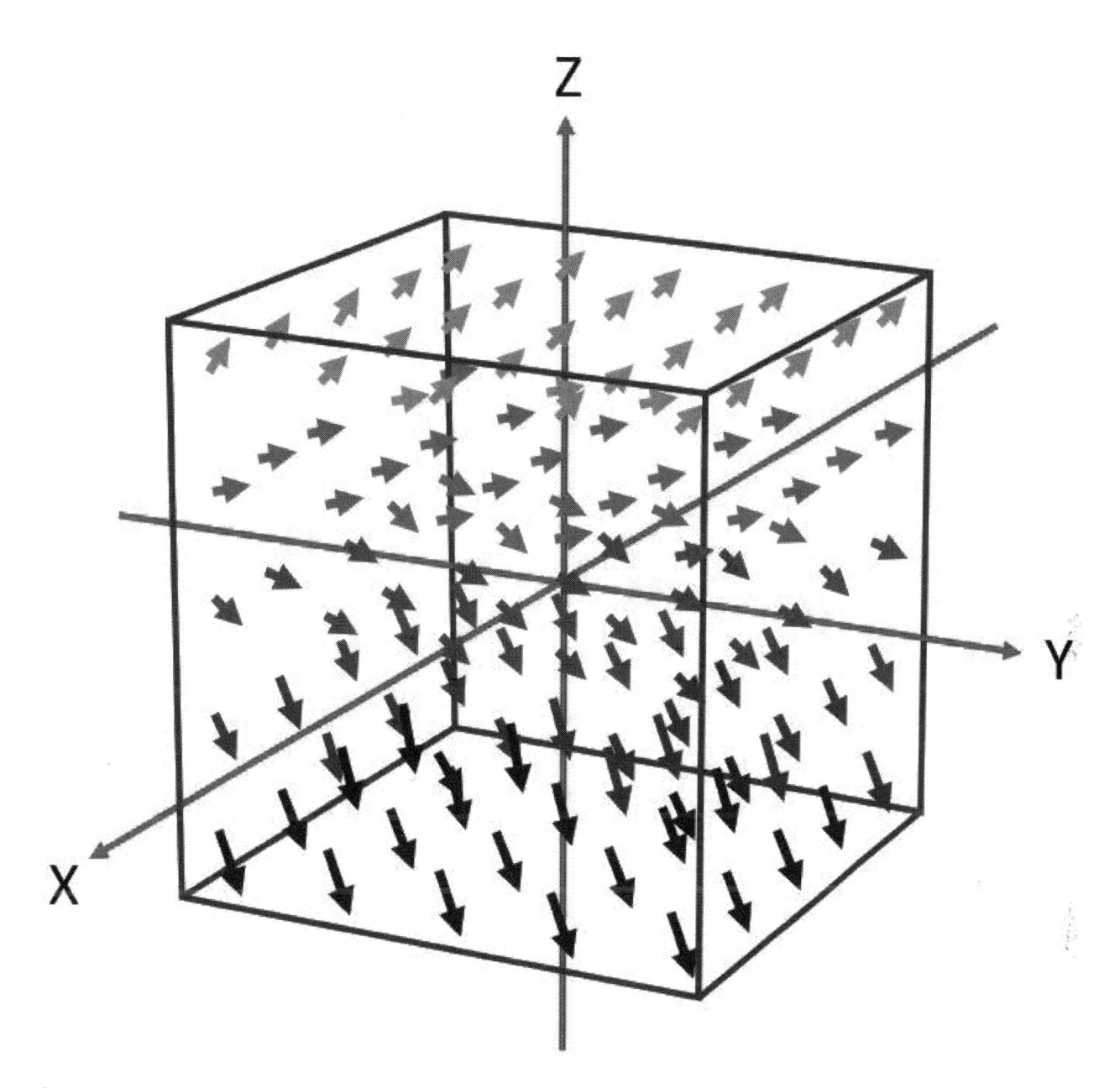

In general, a vector field **F(x, y, z)** can be expressed as,

$$\vec{F}(x, y, z) = F_x(x, y, z)\hat{i} + F_y(x, y, z)\hat{j} + F_z(x, y, z)\hat{k}$$

where the components **F_x**, **F_y** and **F_z** are scalar fields.

Solved Examples:

3. Find the unit vector at a point P(2,3,4) in the vector field F= x $\hat{i}$ + (x²+y² -z) $\hat{j}$ + z $\hat{k}$.

$$\vec{F} = x\hat{\imath} + (x^2 + y^2 - z)\hat{\jmath} + z\hat{k}$$

$$\vec{F}_{P(2,3,4)} = 2\hat{\imath} + 9\hat{\jmath} + 4\hat{k}$$

$$\text{Unit Vector} = \frac{2\hat{\imath} + 9\hat{\jmath} + 4\hat{k}}{\sqrt{2^2 + 9^2 + 4^2}}$$

$$= \frac{2\hat{\imath} + 9\hat{\jmath} + 4\hat{k}}{\sqrt{101}}$$

3.6 PARTIAL DERIVATIVE & THE ∇ OPERATOR

For single variable functions, the derivative measures the rate of change of the function with respect to the independent variable. But for multi variable functions (like a scalar field), the function can vary with respect to more than one variable. For example, the function **F = xyz³** varies with either of **x, y** or **z**. For such functions, we use the partial derivative to measure the rate of change of the function with respect to any one of these variables.

The partial derivative is basically the derivative of the function with respect to any one of its variables with the other variables held constant. Partial derivative of a function **f** with respect to **x, y** & **z** are denoted as $\frac{\partial f}{\partial x}$, $\frac{\partial f}{\partial y}$ & $\frac{\partial f}{\partial z}$ respectively. By considering the other variables as constant, we are essentially reducing the multi variable function to a single variable function and thereby measuring the rate of change of the function with respect to any one variable.

Consider the graph shown below. Here the function **f(x,y)** varies with both **x** and **y**. Now, by keeping the **x** variable as constant (figure 1), we have obtained the variation of the function with respect to **y** alone (figure 2).

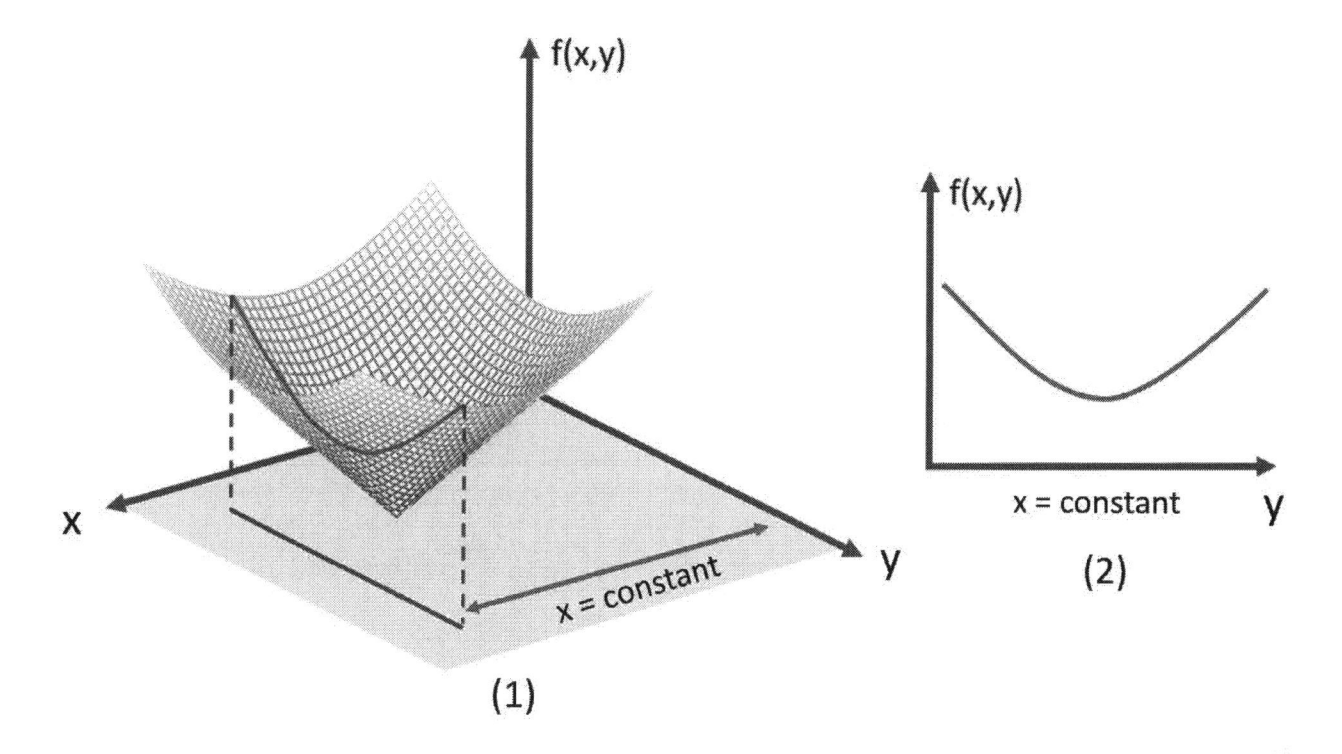

In the above example we have used a 2-variable function for ease of visualization, but whatever's said here is directly applicable to a 3-variable or higher order functions.

Using partial derivatives, we can define a new operator called the del or the nabla operator, denoted by the symbol ∇. In Cartesian coordinates, the del operator is defined as:

$$\nabla = \frac{\partial}{\partial x}\hat{\imath} + \frac{\partial}{\partial y}\hat{\jmath} + \frac{\partial}{\partial z}\hat{k}$$

∇ isn't really a vector, it is rather a differential operator. For a 3 variable function, it can be applied in three different forms (gradient, divergence, curl), each of which is discussed in detail in the next chapter.

Solved Examples:

4. Find the partial derivative of f= $x^2 + y^3$ with respect to x and y.

To find the partial derivative with respect to **x**, variable **y** is treated as a constant. Similarly, to find the partial derivative with respect to **y**, variable **x** is treated as a constant.

$$f = x^2 + y^3$$

$$\frac{\partial f}{\partial x} = 2x, \quad \}y \text{ is constant}$$

$$\frac{\partial f}{\partial y} = 3y^2, \quad \}x \text{ is constant}$$

5. Find the partial derivatives of f= x⁴ − 3xyz

$$f = x^4 - 3xyz$$

$$\frac{\partial f}{\partial x} = 4x^3 - 3yz, \quad \}y, z \text{ are constant}$$

$$\frac{\partial f}{\partial y} = -3xz, \quad \}x, z \text{ are constant}$$

$$\frac{\partial f}{\partial z} = -3xy \quad \}x, y \text{ are constant}$$

6. Find the partial derivative of f= x²y + sin x + cos y.

$$f = x^2y + \sin x + \cos y$$

$$\frac{\partial f}{\partial x} = 2xy + \cos x, \quad \}y \text{ is constant}$$

$$\frac{\partial f}{\partial y} = x^2 - \sin y \quad \}x \text{ is constant}$$

4. GRADIENT, DIVERGENCE & CURL

4.1 GRADIENT

As mentioned at the end of the last chapter, a multi variable function has multiple partial derivatives at every point, one with respect to all of its variables (shown in the figure below). Each of these partial derivatives correspond to the rate of change of the function with respect to that particular variable. Now, if we consider these partial derivatives as vectors, then the resultant of these partial derivative vectors denotes the maximum rate of change of the function when all the variables are considered at once. That is the gradient.

In other words, the Gradient of a multi variable function is a vector that points in the direction of greatest increase (steepest slope) of the function at a point. It is denoted by symbol $\nabla \mathbf{f}$. The gradient is analogous to the slope for single variable functions.

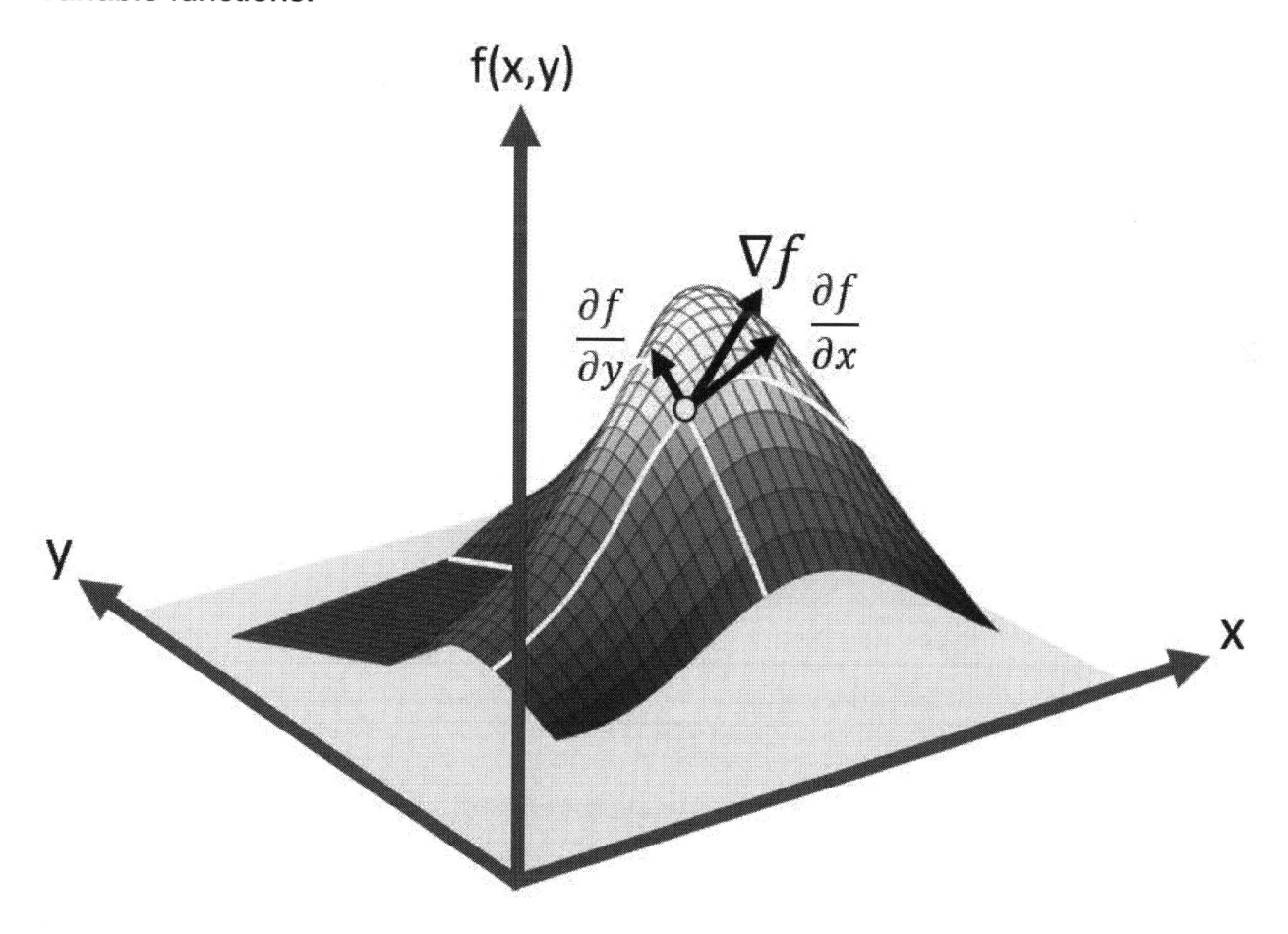

The significance of the gradient can be understood with the help of an example. Consider a mountain whose height above the sea level at point **(x,y)** is given by the scalar field **H(x,y)**. Then the gradient of H at a point will be in the direction of the steepest slope at that point. So if you want to climb down the mountain along the shortest path, all you have to do is to keep following the gradients (direction opposite to the gradients), starting with your current position.

Similarly, if V is the potential due to an electric charge, then the force acting on a unit charge at the point (x, y, z) is in the direction of most rapid decrease of the potential. The gradient has many such applications in physics and engineering.

In another way, the gradient can be thought of as an operator that converts a scalar field into a vector field.

Scalar field f → Gradient Vector field ∇f

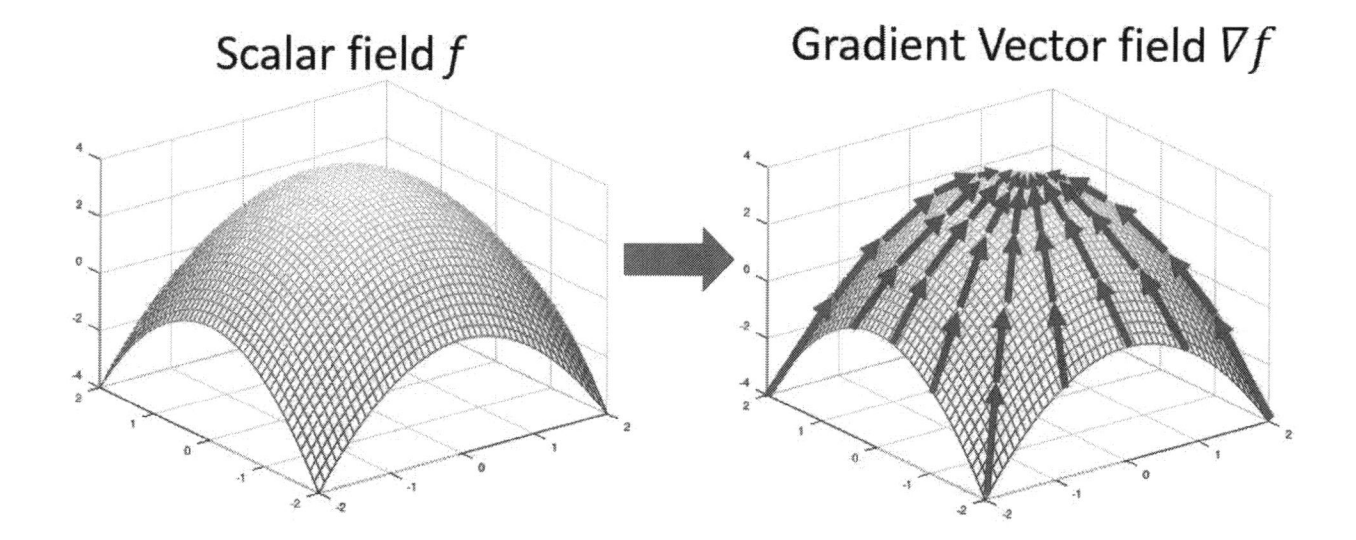

Mathematically the gradient can be obtained as:

$$\nabla f(x, y, z) = (\hat{i}\frac{\partial}{\partial x} + \hat{j}\frac{\partial}{\partial y} + \hat{k}\frac{\partial}{\partial z})f$$

$$= (\hat{i}\frac{\partial f}{\partial x} + \hat{j}\frac{\partial f}{\partial y} + \hat{k}\frac{\partial f}{\partial z})$$

Directional derivative:

The slope or the rate of change of a function in any random direction (not just along the x or the y axis) is called the directional derivative. Finding the directional derivative along a random direction is not as easy as finding the partial derivative. In practice the directional derivative in the direction of the unit vector $\vec{u}$ can be obtained using the gradient ∇f as,

$$D_{\vec{u}}f = \nabla f \cdot \vec{u}$$

Properties of the Gradient:

- $\nabla(u + v) = \nabla u + \nabla v$

$$\nabla(u + v) = \left(\frac{\partial}{\partial x}\hat{\imath} + \frac{\partial}{\partial y}\hat{\jmath} + \frac{\partial}{\partial z}\hat{k}\right)(u + v)$$

$$= \left(\frac{\partial u}{\partial x}\hat{\imath} + \frac{\partial u}{\partial y}\hat{\jmath} + \frac{\partial u}{\partial z}\hat{k}\right) + \left(\frac{\partial v}{\partial x}\hat{\imath} + \frac{\partial v}{\partial y}\hat{\jmath} + \frac{\partial v}{\partial z}\hat{k}\right)$$

$$= \nabla u + \nabla v$$

- $\nabla(uv) = (\nabla u)v + (\nabla v)u$

$$\nabla(uv) = \left(\frac{\partial}{\partial x}\hat{\imath} + \frac{\partial}{\partial y}\hat{\jmath} + \frac{\partial}{\partial z}\hat{k}\right)(uv)$$

$$= \frac{\partial(uv)}{\partial x}\hat{\imath} + \frac{\partial(uv)}{\partial y}\hat{\jmath} + \frac{\partial(uv)}{\partial z}\hat{k}$$

Using Product rule,

$$\frac{\partial(uv)}{\partial x} = u\frac{\partial(v)}{\partial x} + v\frac{\partial(u)}{\partial x}$$

$$\therefore \nabla(uv) = (u\frac{\partial(v)}{\partial x} + v\frac{\partial(u)}{\partial x})\hat{\imath} + (u\frac{\partial(v)}{\partial y} + v\frac{\partial(u)}{\partial y})\hat{\jmath}$$
$$+ (u\frac{\partial(v)}{\partial y} + v\frac{\partial(u)}{\partial y})\hat{k}$$

$$= u(\frac{\partial(v)}{\partial x}\hat{\imath} + \frac{\partial(v)}{\partial y}\hat{\jmath} + \frac{\partial(v)}{\partial z}\hat{k})$$
$$+ v(\frac{\partial(u)}{\partial x}\hat{\imath} + \frac{\partial(u)}{\partial y}\hat{\jmath} + \frac{\partial(u)}{\partial z}\hat{k})$$
$$= (\nabla u)v + (\nabla v)u$$

- **k ∇(u) = ∇(ku)**, where **k** is a scalar quantity

Solved Examples:

1. **Find the gradient of the scalar field f = 2x + yz at point P(5,-3, 9)**

$$\nabla f = \frac{\partial(2x+yz)}{\partial x}\hat{\imath} + \frac{\partial(2x+yz)}{\partial y}\hat{\jmath} + \frac{\partial(2x+yz)}{\partial z}\hat{k}$$
$$= 2\hat{\imath} + z\hat{\jmath} + y\hat{k}$$

At point P(5,-3, 9), the gradient is denoted by the vector $2\hat{\imath} + 9\hat{\jmath} - 3\hat{k}$ & the magnitude of the gradient is **9.695**.

2. **Find the normal vector to the surface $x^3+y^3z = 3$ at the point (1,1,2).**

Equation of a surface is **f(x,y,z) = c** and a normal vector on the surface is given by ∇**f(x,y,z)**.

$$\nabla f = \frac{\partial(x^3 + y^3z)}{\partial x}\hat{\imath} + \frac{\partial(x^3 + y^3z)}{\partial y}\hat{\jmath} + \frac{\partial(x^3 + y^3z)}{\partial z}\hat{k}$$

$$= 3x^2\,\hat{\imath} + 3y^2z\,\hat{\jmath} + y^3\,\hat{k}$$

At point (1,1,2):

Normal vector: $3\hat{\imath} + 6\hat{\jmath} + \hat{k}$

3. Find the directional derivative of f(x,y)=x²y in the direction of (1,2) at the point (3,2).

$$\nabla f = \frac{\partial(x^2y)}{\partial x}\hat{\imath} + \frac{\partial(x^2y)}{\partial y}\hat{\jmath}$$

$$= 2xy\,\hat{\imath} + x^2\,\hat{\jmath}$$

$$\nabla f(3,2) = 12\hat{\imath} + 9\hat{\jmath}$$

Unit vector in the direction of $\hat{\imath} + 2\hat{\jmath}$

$$\hat{u} = \frac{\hat{\imath} + 2\hat{\jmath}}{\sqrt{5}}$$

∴ Directional derivative at point (3,2) in the direction of the unit vector $\hat{u}$

$$= \nabla f(3,2) \cdot \hat{u}$$

$$= (12\hat{\imath} + 9\hat{\jmath}) \cdot \left(\frac{\hat{\imath} + 2\hat{\jmath}}{\sqrt{5}}\right)$$

$$= \frac{12 + 18}{\sqrt{5}} = \frac{30}{\sqrt{5}}$$

4.2 DIVERGENCE

We saw how the del operator can be used on a scalar function in the form of the gradient. On a vector field the del operator can be used in 2 ways, first of which is the Divergence.

The divergence is nothing but the dot product between the del operator and a vector field.

If $V = V_x \hat{i} + V_y \hat{j} + V_z \hat{k}$,

$$\nabla \cdot V = \left(\frac{\partial}{\partial x}\hat{i} + \frac{\partial}{\partial y}\hat{j} + \frac{\partial}{\partial z}\hat{k}\right) \cdot (V_x \hat{i} + V_y \hat{j} + V_z \hat{k})$$

$$= \frac{\partial V_x}{\partial x} + \frac{\partial V_y}{\partial y} + \frac{\partial V_z}{\partial z}$$

Intuitively, the Divergence represents the outward flow of a vector field from an infinitesimal volume at a given point in the field. In other words, divergence is a measure of the extent to which a point (which is essentially a tiny volume) behaves as a source of the vector field.

To understand the concept of divergence better, imagine the vector field as a fluid flow as shown below.

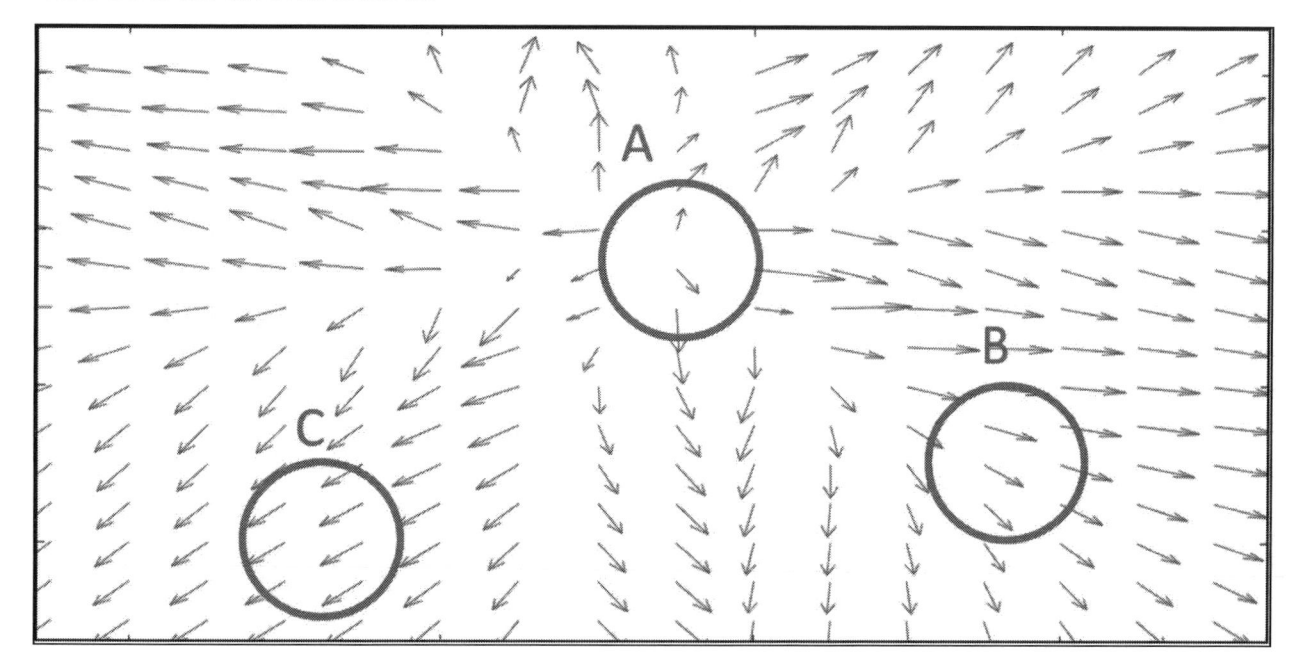

Now if you consider a small spherical volume, the difference between the outward flow and the inward flow i.e. the net outward flow gives the divergence of the flow in the small volume. For example, at point A, all the field lines are pointed away from the volume, which means point A is acting as source of the flux, therefore the divergence at that point is a positive value. At point B, some of the field lines are flowing into the volume and some are flowing out of the volume, but because there are more outward flowing field lines, the net outward flow is positive and therefore divergence is positive at point B as well (but it has less magnitude compared to point A). At point C, there are equal no. of field lines flowing into the volume as there are field lines flowing out of the volume. Hence the divergence at point C is zero.

The above example is only for a better understanding of the concept, in reality the divergence has nothing to do with the no. of field lines entering or exiting the volume, it has more to do with the magnitude & direction of the field lines.

Just as the gradient operator converts a scalar field into a vector field, the divergence operator does the opposite, it converts a vector field into a scalar field.

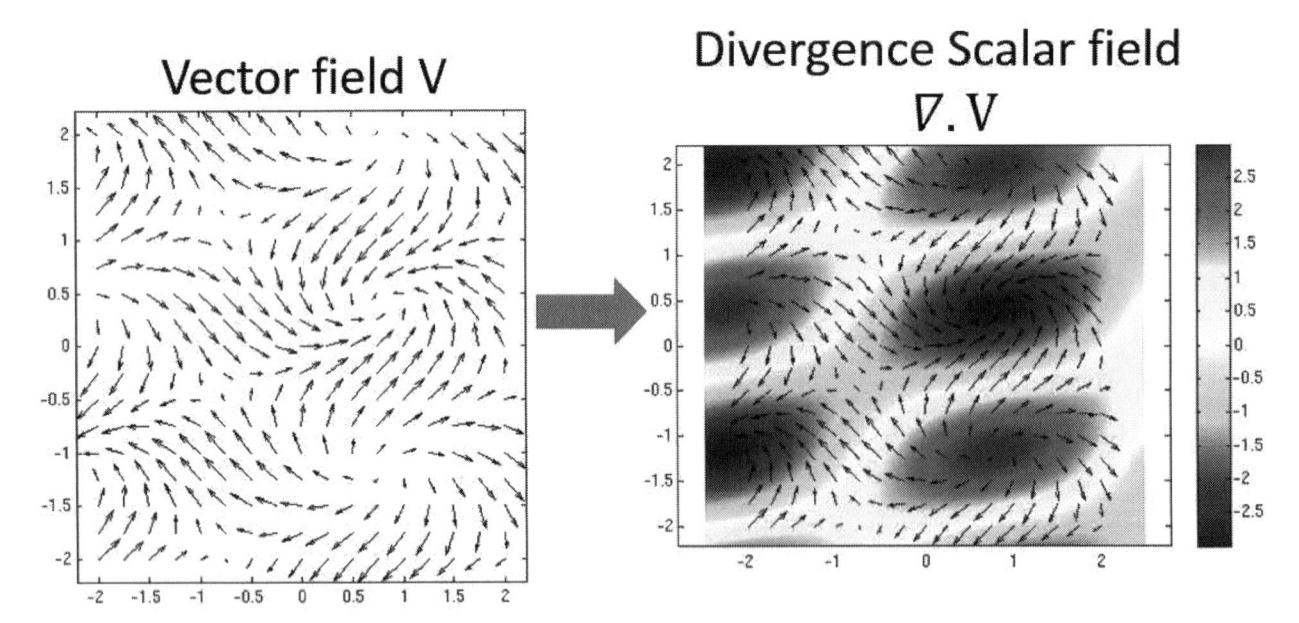

Vector field V

Divergence Scalar field $\nabla . V$

Properties of Divergence:

- $\nabla . (A + B) = \nabla . A + \nabla . B$

$$\vec{A} = a_1 \hat{i} + a_2 \hat{j} + a_3 \hat{k}$$
$$\vec{B} = b_1 \hat{i} + b_2 \hat{j} + b_3 \hat{k}$$
$$\therefore \vec{A} + \vec{B} = (a_1 + b_1) \hat{i} + (a_2 + b_2) \hat{j} + (a_3 + b_3) \hat{k}$$

$$\nabla \cdot (\vec{A} + \vec{B}) =$$
$$\left(\frac{\partial}{\partial x} \hat{i} + \frac{\partial}{\partial y} \hat{j} + \frac{\partial}{\partial z} \hat{k}\right) \cdot \left((a_1 + b_1) \hat{i} + (a_2 + b_2) \hat{j} + (a_3 + b_3) \hat{k}\right)$$
$$= \frac{\partial (a_1 + b_1)}{\partial x} + \frac{\partial (a_2 + b_2)}{\partial y} + \frac{\partial (a_3 + b_3)}{\partial z}$$
$$= \nabla \cdot \vec{A} + \nabla \cdot \vec{B}$$

- $\nabla \cdot (f\vec{A}) = f \nabla \cdot \vec{A} + \vec{A} \cdot \nabla f$, where *f* is a scalar function

$$\vec{A} = a_1 \hat{i} + a_2 \hat{j} + a_3 \hat{k}$$

$$\nabla \cdot (f\vec{A}) = \left(\frac{\partial}{\partial x}\hat{i} + \frac{\partial}{\partial y}\hat{j} + \frac{\partial}{\partial z}\hat{k}\right) \cdot (fa_1\hat{i} + fa_2\hat{j} + fa_3\hat{k})$$

$$= \frac{\partial(fa_1)}{\partial x} + \frac{\partial(fa_2)}{\partial y} + \frac{\partial(fa_3)}{\partial z}$$

Using Product rule,

$$\frac{\partial(fa_1)}{\partial x} = f\frac{\partial(a_1)}{\partial x} + a_1\frac{\partial(f)}{\partial x}$$

$$\therefore \nabla \cdot (f\vec{A}) = f\frac{\partial(a_1)}{\partial x} + a_1\frac{\partial(f)}{\partial x} + f\frac{\partial(a_2)}{\partial y} + a_2\frac{\partial(f)}{\partial y}$$
$$+ f\frac{\partial(a_3)}{\partial z} + a_3\frac{\partial(f)}{\partial z}$$

$$= f\left(\frac{\partial(a_1)}{\partial x} + \frac{\partial(a_2)}{\partial y} + \frac{\partial(a_3)}{\partial z}\right) + \left(a_1\frac{\partial f}{\partial x} + a_2\frac{\partial f}{\partial y} + a_3\frac{\partial f}{\partial z}\right)$$

$$= f(\nabla \cdot A) + A \cdot \nabla f$$

Solved Examples:

4. Find the divergence of vector field $V = x\hat{i} + yz\hat{j} + 3xz\hat{k}$

$$\nabla \cdot \vec{V} = (\frac{\partial}{\partial x}\hat{i} + \frac{\partial}{\partial y}\hat{j} + \frac{\partial}{\partial z}\hat{k}) \cdot (x\hat{i} + yz\hat{j} + 3xz\hat{k})$$

$$= \frac{\partial(x)}{\partial x} + \frac{\partial(yz)}{\partial y} + \frac{\partial(3xz)}{\partial z}$$

$$= 1 + z + 3x$$

5. **Compute the divergence of the vector field $F = x^2y\,\hat{i} + xyz\,\hat{j} - x^2y^2\,\hat{k}$ at points P(1,2,-5) and Q(5,2,1). Which of these points act as a source of the vector field?**

$$\nabla \cdot \vec{F} = (\frac{\partial}{\partial x}\hat{i} + \frac{\partial}{\partial y}\hat{j} + \frac{\partial}{\partial z}\hat{k}) \cdot (x^2y\,\hat{i} + xyz\,\hat{j} - x^2y^2\,\hat{k})$$

$$= 2xy + xz$$

At point P(1,2,-5):

$$\nabla \cdot \vec{F} = -1$$

At point Q(5,2,1):

$$\nabla \cdot \vec{F} = 25$$

The divergence at point P is negative, indicating it is acting as a sink for the vector field. Point Q on the contrary has positive divergence and hence it acts as a source of the vector field.

4.3 CURL

The cross product between the del operator and a vector field is called the Curl of a vector field. It is denoted as $\nabla \times \mathbf{V}$. It can be calculated as,

If $V = V_x \hat{i} + V_y \hat{j} + V_z \hat{k}$,

$$\nabla \times V = (\frac{\partial}{\partial x}\hat{i} + \frac{\partial}{\partial y}\hat{j} + \frac{\partial}{\partial z}\hat{k}) \times (V_x \hat{i} + V_y \hat{j} + V_z \hat{k})$$

$$= \begin{vmatrix} \hat{i} & \hat{j} & \hat{k} \\ \frac{\partial}{\partial x} & \frac{\partial}{\partial y} & \frac{\partial}{\partial z} \\ V_x & V_y & V_z \end{vmatrix}$$

$$= (\frac{\partial V_z}{\partial y} - \frac{\partial V_y}{\partial z})\hat{i} - (\frac{\partial V_z}{\partial x} - \frac{\partial V_x}{\partial z})\hat{j} + (\frac{\partial V_y}{\partial x} - \frac{\partial V_x}{\partial y})\hat{k}$$

The curl of a vector field describes the rotational tendency of a vector field at a point in 3d space. Consider a fluid flow as shown in the figure below. Now if you consider a small spherical ball that is free is to rotate in any direction, it will rotate differently depending on its location in the fluid. The field vectors acting on the sphere determines both direction and the speed at which it rotates. Magnitude of the curl vector denotes the speed of rotation and the direction of the curl denotes the axis of rotation of the sphere.

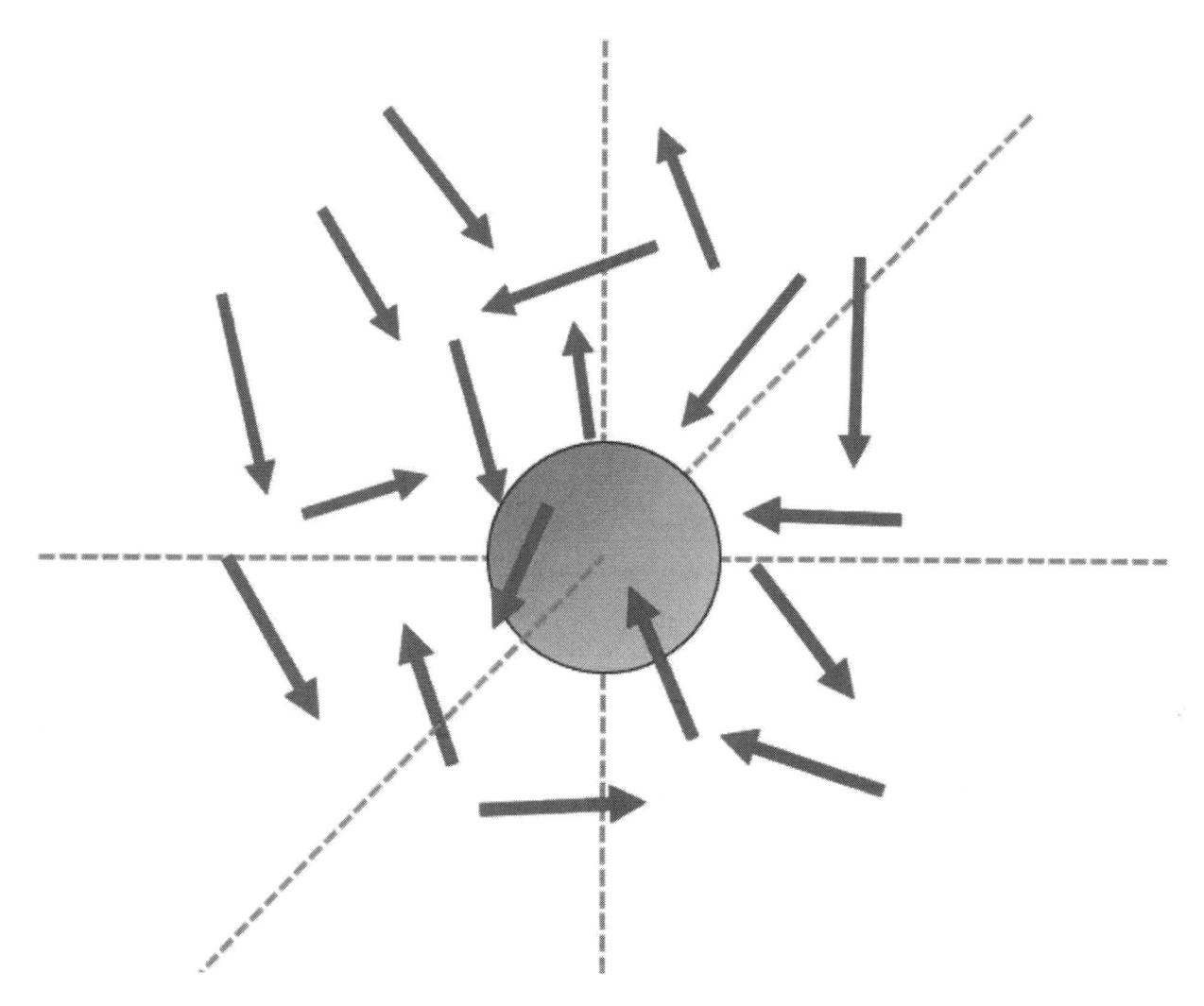

It is important to note that the curl refers to the microscopic rotation of the ball at a point in the vector field (i.e. as if translational motion of the ball from the point is restricted) and not the macroscopic circulation of the ball in the field, if any. If you think about the motion of the earth around the sun, the curl is analogous to the rotation of the earth and not its revolution around the sun.

The Curl operator converts a vector field into another vector field.

Properties of Curl:

- $\nabla \times (A + B) = (\nabla \times A) + (\nabla \times B)$

$$\vec{A} = a_1 \hat{i} + a_2 \hat{j} + a_3 \hat{k}$$
$$\vec{B} = b_1 \hat{i} + b_2 \hat{j} + b_3 \hat{k}$$
$$\vec{A} + \vec{B} = (a_1 + b_1)\hat{i} + (a_2 + b_2)\hat{j} + (a_3 + b_3)\hat{k}$$

$\nabla \times (\vec{A} + \vec{B}) =$
$$\left(\frac{\partial(a_3 + b_3)}{\partial y} - \frac{\partial(a_2 + b_2)}{\partial z}\right)\hat{i} - \left(\frac{\partial(a_3 + b_3)}{\partial x} - \frac{\partial(a_1 + b_1)}{\partial z}\right)\hat{j} + \left(\frac{\partial(a_2 + b_2)}{\partial x} - \frac{\partial(a_1 + b_1)}{\partial y}\right)\hat{k}$$

$$= \left(\frac{\partial(a_3)}{\partial y} - \frac{\partial(a_2)}{\partial z}\right)\hat{i} - \left(\frac{\partial(a_3)}{\partial x} - \frac{\partial(a_1)}{\partial z}\right)\hat{j} + \left(\frac{\partial(a_2)}{\partial x} - \frac{\partial(a_1)}{\partial y}\right)\hat{k}$$
$$+ \left(\frac{\partial(b_3)}{\partial y} - \frac{\partial(b_2)}{\partial z}\right)\hat{i} - \left(\frac{\partial(b_3)}{\partial x} - \frac{\partial(b_1)}{\partial z}\right)\hat{j} + \left(\frac{\partial(b_2)}{\partial x} - \frac{\partial(b_1)}{\partial y}\right)\hat{k}$$

$$= (\nabla \times \vec{A}) + (\nabla \times \vec{B})$$

- $\nabla \times (f\mathbf{A}) = f(\nabla \times \mathbf{A}) + (\nabla f \times \mathbf{A})$, where f is a scalar function

The property can be proven in the same way as we did in the above case. Use the product to expand the derivative.

Solved Examples:

6. Find the curl of the vector field $V = x\hat{i} + yz\hat{j} + 3xz\hat{k}$

$$\nabla \times \vec{V} = \begin{vmatrix} \hat{i} & \hat{j} & \hat{k} \\ \dfrac{\partial}{\partial x} & \dfrac{\partial}{\partial y} & \dfrac{\partial}{\partial z} \\ x & yz & 3xz \end{vmatrix}$$

$$= \left(\dfrac{\partial(3xz)}{\partial y} - \dfrac{\partial(yz)}{\partial z}\right)\hat{i} - \left(\dfrac{\partial(3xz)}{\partial x} - \dfrac{\partial(x)}{\partial z}\right)\hat{j}$$
$$+ \left(\dfrac{\partial(yz)}{\partial x} - \dfrac{\partial(x)}{\partial y}\right)\hat{k}$$

$$= -y\hat{i} - 3z\hat{j}$$

7. Compute the curl of the vector field $F = x^2 y\,\hat{i} + xyz\,\hat{j} - x^2 y^2\,\hat{k}$ at point P(1,2,-5).

$$\nabla \times \vec{F} = \begin{vmatrix} \hat{i} & \hat{j} & \hat{k} \\ \dfrac{\partial}{\partial x} & \dfrac{\partial}{\partial y} & \dfrac{\partial}{\partial z} \\ x^2 y & xyz & -x^2 y^2 \end{vmatrix}$$

$$= \left(\dfrac{\partial(-x^2 y^2)}{\partial y} - \dfrac{\partial(xyz)}{\partial z}\right)\hat{i} - \left(\dfrac{\partial(-x^2 y^2)}{\partial x} - \dfrac{\partial(x^2 y)}{\partial z}\right)\hat{j} + \left(\dfrac{\partial(xyz)}{\partial x} - \dfrac{\partial(x^2 y)}{\partial y}\right)\hat{k}$$

$$= (-2x^2 y - xy)\hat{i} + (2xy^2)\hat{j} + (yz - x^2)\hat{k}$$

At point P(1,2,-5):

$$\nabla \times \vec{F} = -6\hat{i} + 8\hat{j} - 11\hat{k}$$

4.4 MORE PROPERTIES

- Divergence of Curl of a vector field is zero i.e. $\nabla \cdot (\nabla \times A) = 0$

$$\vec{A} = a_1 \hat{i} + a_2 \hat{j} + a_3 \hat{k}$$

$\nabla \times \vec{A} =$

$$\left(\frac{\partial(a_3)}{\partial y} - \frac{\partial(a_2)}{\partial z}\right)\hat{i} - \left(\frac{\partial(a_3)}{\partial x} - \frac{\partial(a_1)}{\partial z}\right)\hat{j} + \left(\frac{\partial(a_2)}{\partial x} - \frac{\partial(a_1)}{\partial y}\right)\hat{k}$$

$$\nabla \cdot (\nabla \times \vec{A}) = \left(\frac{\partial}{\partial x}\hat{i} + \frac{\partial}{\partial y}\hat{j} + \frac{\partial}{\partial z}\hat{k}\right) \cdot (\nabla \times \vec{A})$$

$$= \frac{\partial^2(a_3)}{\partial x \partial y} - \frac{\partial^2(a_2)}{\partial x \partial z} - \frac{\partial^2(a_3)}{\partial x \partial y} + \frac{\partial^2(a_1)}{\partial y \partial z} + \frac{\partial^2(a_2)}{\partial z \partial x} - \frac{\partial^2(a_1)}{\partial y \partial z}$$

$$\Rightarrow \boxed{\nabla \cdot (\nabla \times \vec{A}) = 0}$$

- Curl of Gradient of a scalar field is zero i.e. $\nabla \times (\nabla f) = 0$

$$\nabla f = \frac{\partial f}{\partial x}\hat{i} + \frac{\partial f}{\partial y}\hat{j} + \frac{\partial f}{\partial z}\hat{k}$$

$\nabla \times (\nabla f) =$

$$\left(\frac{\partial^2 f}{\partial y \partial z} - \frac{\partial^2 f}{\partial z \partial y}\right)\hat{i} - \left(\frac{\partial^2 f}{\partial x \partial z} - \frac{\partial^2 f}{\partial z \partial x}\right)\hat{j} + \left(\frac{\partial^2 f}{\partial x \partial y} - \frac{\partial^2 f}{\partial y \partial x}\right)\hat{k}$$

$$\Rightarrow \boxed{\nabla \times (\nabla f) = 0}$$

- $\nabla \cdot (A \times B) = B \cdot (\nabla \times A) - A \cdot (\nabla \times B)$

$$\vec{A} = a_1 \hat{i} + a_2 \hat{j} + a_3 \hat{k}$$
$$\vec{B} = b_1 \hat{i} + b_2 \hat{j} + b_3 \hat{k}$$

$$\vec{A} \times \vec{B} = (a_2 b_3 - a_3 b_2)\hat{i} + (a_3 b_1 - a_1 b_3)\hat{j} + (a_1 b_2 - a_2 b_1)\hat{k}$$

$$\therefore \nabla \cdot (\vec{A} \times \vec{B}) = \frac{\partial(a_2 b_3 - a_3 b_2)}{\partial x} + \frac{\partial(a_3 b_1 - a_1 b_3)}{\partial y} + \frac{\partial(a_1 b_2 - a_2 b_1)}{\partial z}$$

$$= \text{RHS}$$

$$\nabla \times \vec{A} =$$
$$\left(\frac{\partial(a_3)}{\partial y} - \frac{\partial(a_2)}{\partial z}\right)\hat{i} + \left(\frac{\partial(a_1)}{\partial z} - \frac{\partial(a_3)}{\partial x}\right)\hat{j} + \left(\frac{\partial(a_2)}{\partial x} - \frac{\partial(a_1)}{\partial y}\right)\hat{k}$$

$$\therefore \vec{B}\cdot(\nabla \times \vec{A}) = b_1\left(\frac{\partial(a_3)}{\partial y} - \frac{\partial(a_2)}{\partial z}\right) + b_2\left(\frac{\partial(a_1)}{\partial z} - \frac{\partial(a_3)}{\partial x}\right)$$
$$+ b_3\left(\frac{\partial(a_2)}{\partial x} - \frac{\partial(a_1)}{\partial y}\right)$$

Similarly,
$$\vec{A}\cdot(\nabla \times \vec{B}) = a_1\left(\frac{\partial(b_3)}{\partial y} - \frac{\partial(b_2)}{\partial z}\right) + a_2\left(\frac{\partial(b_1)}{\partial z} - \frac{\partial(b_3)}{\partial x}\right)$$
$$+ a_3\left(\frac{\partial(b_2)}{\partial x} - \frac{\partial(b_1)}{\partial y}\right)$$

$$\therefore \vec{B}\cdot(\nabla \times \vec{A}) - \vec{A}\cdot(\nabla \times \vec{B}) =$$
$$\frac{\partial(a_2 b_3 - a_3 b_2)}{\partial x} + \frac{\partial(a_3 b_1 - a_1 b_3)}{\partial y} + \frac{\partial(a_1 b_2 - a_2 b_1)}{\partial z}$$

$$= \text{LHS}$$

Comparing the LHS and the RHS we can see both the terms are equal and hence the relation is proved.

- $\nabla \times (\nabla \times A) = \nabla(\nabla \cdot A) - \nabla^2 A$

Proof of this identity can be obtained the same way as in above case. Try it out.

Solved Examples:

8. Using the scalar field f = xy + yz, prove that the curl of the gradient of a scalar field is zero.

$$f = xy + yz$$

$$\nabla f = \frac{\partial(xy+yz)}{\partial x}\hat{\imath} + \frac{\partial(xy+yz)}{\partial y}\hat{\jmath} + \frac{\partial(xy+yz)}{\partial z}\hat{k}$$

$$= y\hat{\imath} + (x+z)\hat{\jmath} + y\hat{k}$$

$$\nabla \times \nabla f = \begin{vmatrix} \hat{\imath} & \hat{\jmath} & \hat{k} \\ \frac{\partial}{\partial x} & \frac{\partial}{\partial y} & \frac{\partial}{\partial z} \\ y & (x+z) & y \end{vmatrix}$$

$$= \left(\frac{\partial(y)}{\partial y} - \frac{\partial(x+z)}{\partial z}\right)\hat{\imath} - \left(\frac{\partial(y)}{\partial x} - \frac{\partial(y)}{\partial z}\right)\hat{\jmath}$$

$$+ \left(\frac{\partial(x+z)}{\partial x} - \frac{\partial(y)}{\partial y}\right)\hat{k}$$

$$= (1-1)\hat{\imath} + (0-0)\hat{\jmath} + (1-1)\hat{k}$$

$$= 0$$

9. Using Vector field F=xy $\hat{\imath}$ + xyz $\hat{\jmath}$ +x^2y^2 $\hat{k}$, prove that the divergence of the curl of the vector field is zero.

$$\vec{F} = xy\,\hat{\imath} + xyz\,\hat{\jmath} + x^2y^2\,\hat{k}$$

$$\nabla \times \vec{F} = \begin{vmatrix} \hat{\imath} & \hat{\jmath} & \hat{k} \\ \dfrac{\partial}{\partial x} & \dfrac{\partial}{\partial y} & \dfrac{\partial}{\partial z} \\ xy & xyz & x^2y^2 \end{vmatrix}$$

$$= \left(\dfrac{\partial(x^2y^2)}{\partial y} - \dfrac{\partial(xyz)}{\partial z}\right)\hat{\imath} - \left(\dfrac{\partial(x^2y^2)}{\partial x} - \dfrac{\partial(xy)}{\partial z}\right)\hat{\jmath}$$

$$+ \left(\dfrac{\partial(xyz)}{\partial x} - \dfrac{\partial(xy)}{\partial y}\right)\hat{k}$$

$$= (2x^2y - xy)\hat{\imath} - 2xy^2\,\hat{\jmath} + (yz - x)\hat{k}$$

$$\nabla \cdot (\nabla \times \vec{F}) = \dfrac{\partial(2x^2y - xy)}{\partial x}\hat{\imath} + \dfrac{\partial(-2xy^2)}{\partial y}\hat{\jmath} + \dfrac{\partial(yz - x)}{\partial z}\hat{k}$$

$$= 4xy - y - 4xy + y = 0$$

4.5 LAPLACIAN

The divergence of the gradient of a scalar field is called the Laplacian, denoted as ∇^2.

$$\nabla^2 f = \nabla \cdot \nabla f$$

The Laplacian is analogous to the second derivative in ordinary calculus. The gradient at a point (as mentioned earlier) denotes a vector that points in the direction of maximum increase of the function at the point. So at the peaks, the gradient vectors are pointed towards it. And at the troughs, the gradient vectors are pointed away from it. (Shown in the figure below)

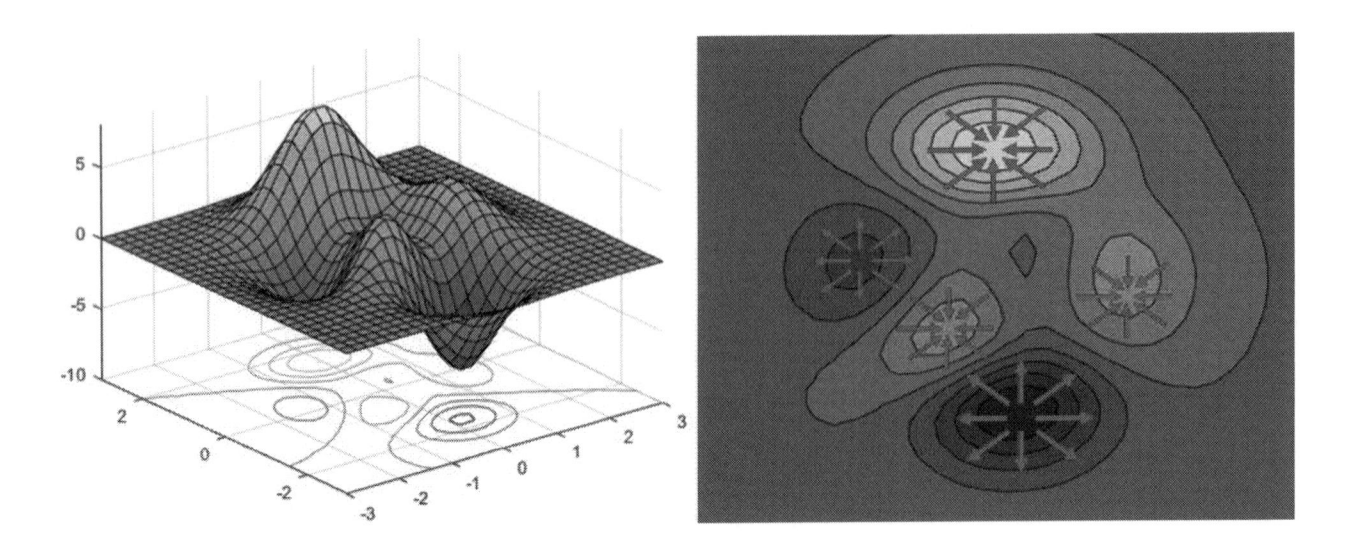

Now, we know that the divergence represents the source like nature at a point. Therefore, at the peaks, the divergence of the gradient is negative maximum and at the troughs, the divergence of the gradient is positive maximum. So the Laplacian of a scalar field tells you how peaky (or trough like) a certain point in the field is.

Solved Examples:

10. Find the Laplacian of the scalar field $f = x^2 + y^3$.

$$\nabla^2 f = \frac{\partial^2 (x^2 + y^3)}{\partial x^2} + \frac{\partial^2 (x^2 + y^3)}{\partial y^2}$$

$$= \frac{\partial (2x)}{\partial x} + \frac{\partial (3y^2)}{\partial y}$$

$$= 2 + 6y$$

5. VECTOR INTEGRATION

In this chapter we discuss the concept of line and surface integrals and a few related theorems.

5.1 LINE INTEGRAL

A) Line Integral of a Scalar field

Line integral or the path integral is the integral of a function evaluated along a curve. For instance, consider a single variable function as shown below, here the line integral along the path AB is the sum of the values of the function **f(x)** as **x** moves from A to B.

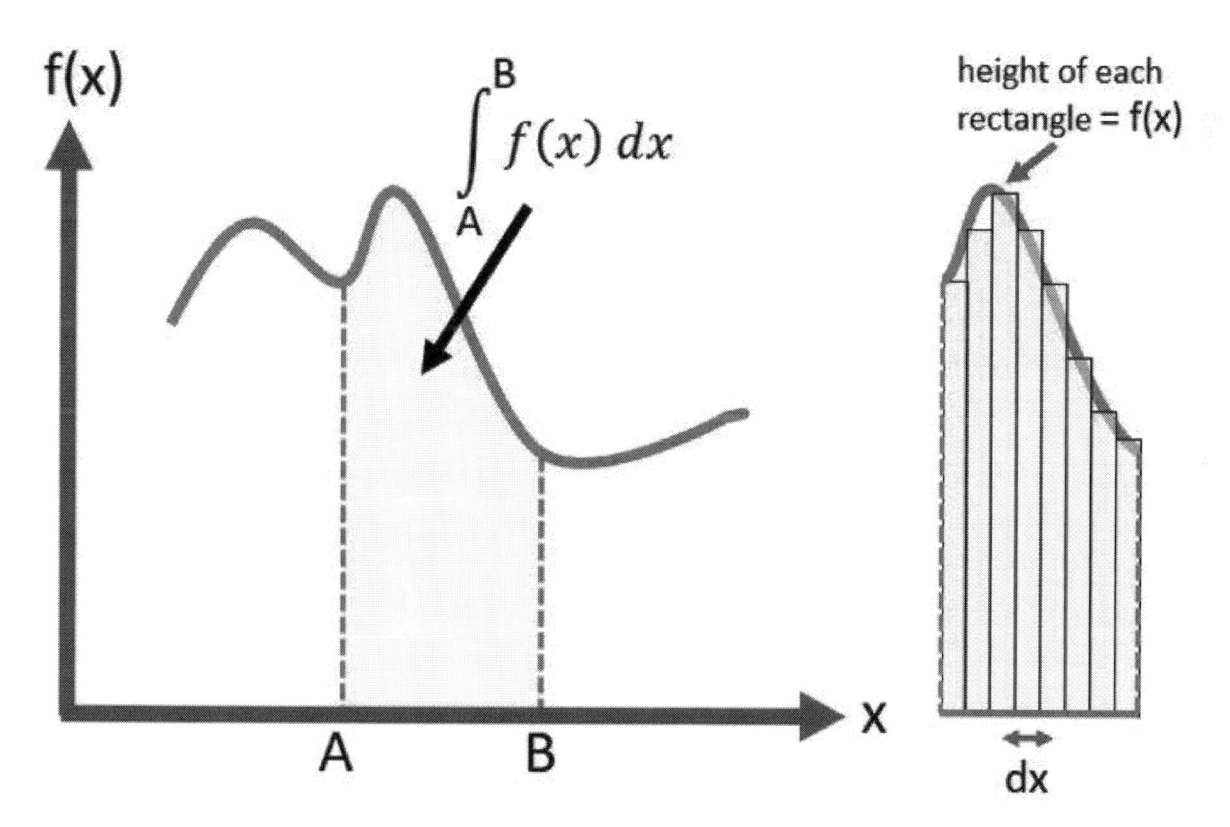

$$\int_A^B f(x)\, dx$$

height of each rectangle = f(x)

The line integral in this case is evaluated by considering a bunch of small rectangular strips of width **dx** and suitable height (which will be equal to the value of the function at the point i.e. **f(x)**), so as to fill up the area. The area of an individual strip is therefore **f(x) dx** (width x height) and hence the area of our interest is simply the integral (summation) of these small areas along the path.

Here things were easy because for a single variable function the line integral can only be evaluated over a straight-line path for obvious reasons. But in case of multivariable functions, the line integral can be evaluated along any 2-dimensional path, not just a straight line.

In the figure shown below, the line integral of the function **f(x,y)** is evaluated along the path AB.

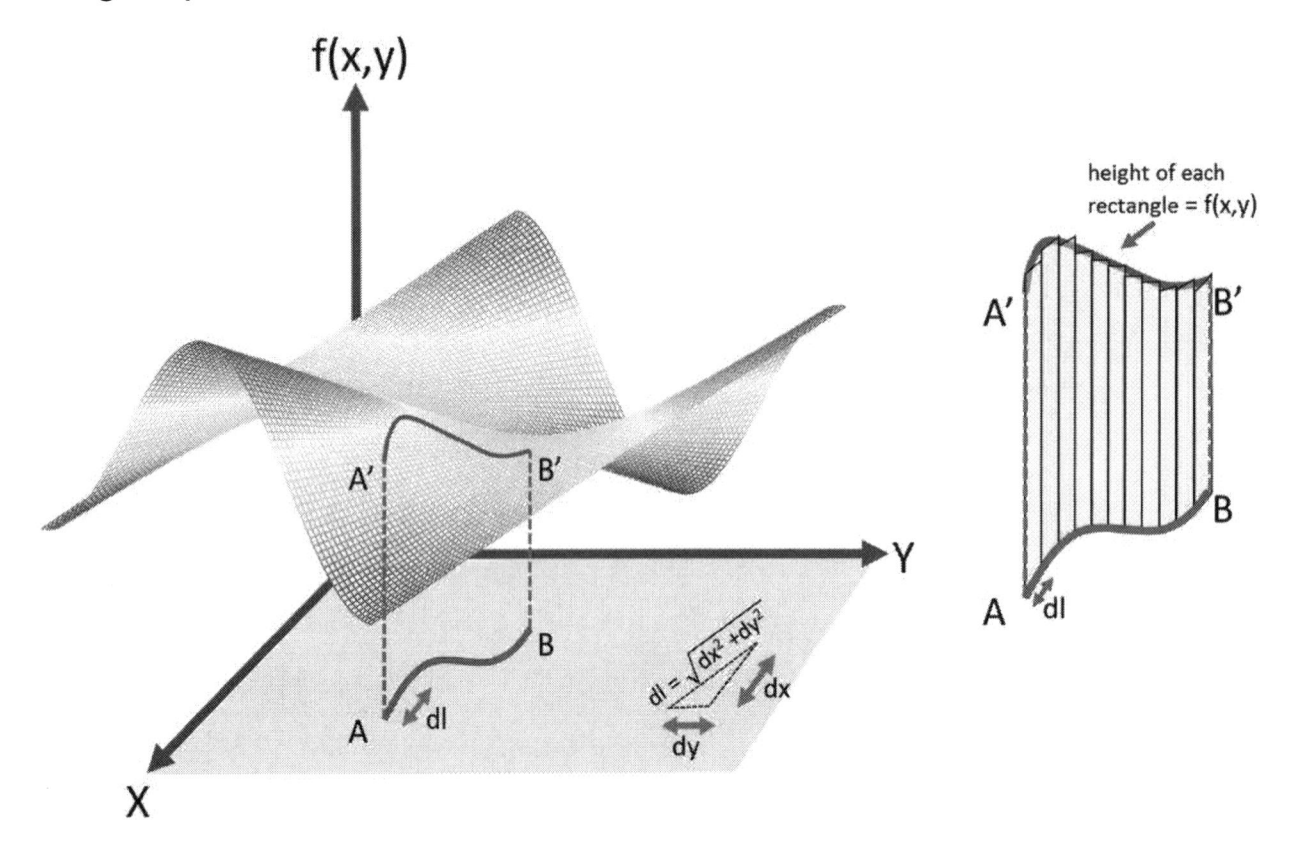

Geometrically, the line integral of a scalar function is nothing but the area of the fence created by a curve path and its projection on the function. And it is evaluated in pretty much the same manner as with single variable functions, except that in this case the line element (width of the rectangular strip) is no longer **dx** or **dy**, but a function of **dx** and **dy**.

In general, the line integral of a function **f** along a curve C is expressed as:

$$\int_C f \, dl$$

$$\int_A^B f(x,y)\,dx \qquad \int_A^B f(x,y)\,dy \qquad \int_A^B f(x,y)\,dl$$

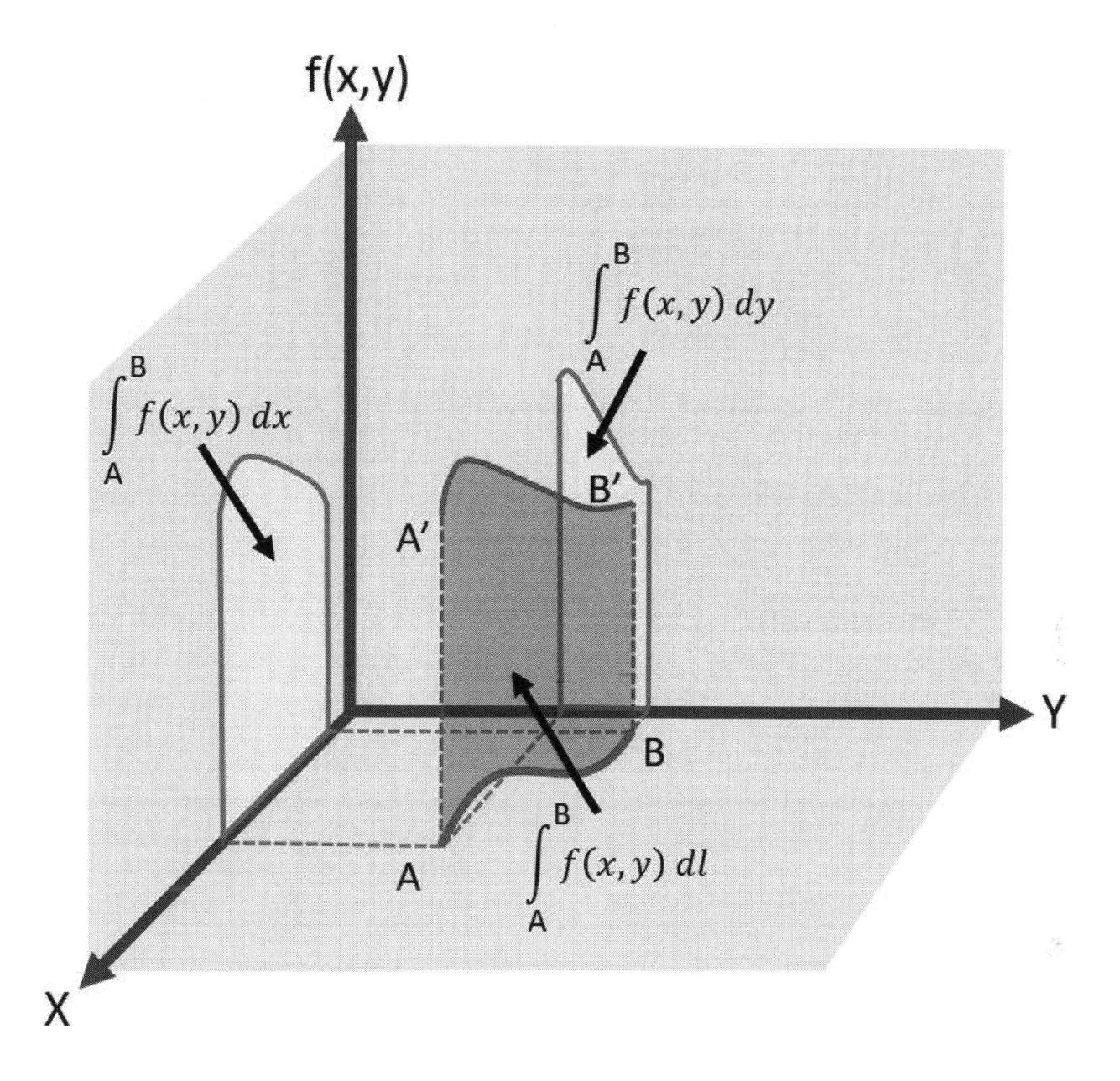

Parameterization of the curve:

We mentioned that for a general curve the line element **dl** is a function of **dx** and **dy** (even **dz** in the case of curves in 3d space). As far as the definition goes that is ok, but when it comes to actually computing the line integral there is a problem. There are 2 variables (**x** & **y**), so how do we integrate it with respect to **dl**? For that we need to do something called Parameterization of the curve, meaning we need to express the curve in terms of a single variable. That doesn't sound right?

Consider a curve AB as shown below and a particle moving along it, whose location is denoted by the coordinates (**x, y**). Now suppose we have a slider indicating time. When the slider is at the left extreme, the particle is at the starting position, then as the slider is moved the location of the particle also changes correspondingly. So basically we have expressed **x** and **y** as functions of time **t**, thereby expressing the whole curve as a function of a single variable **t**. While it is not necessary for the third parameter to be time, but for convenience that is usually the case. Important thing to note is that, for parameterization to work an interval for parameter **t** has to be specified.

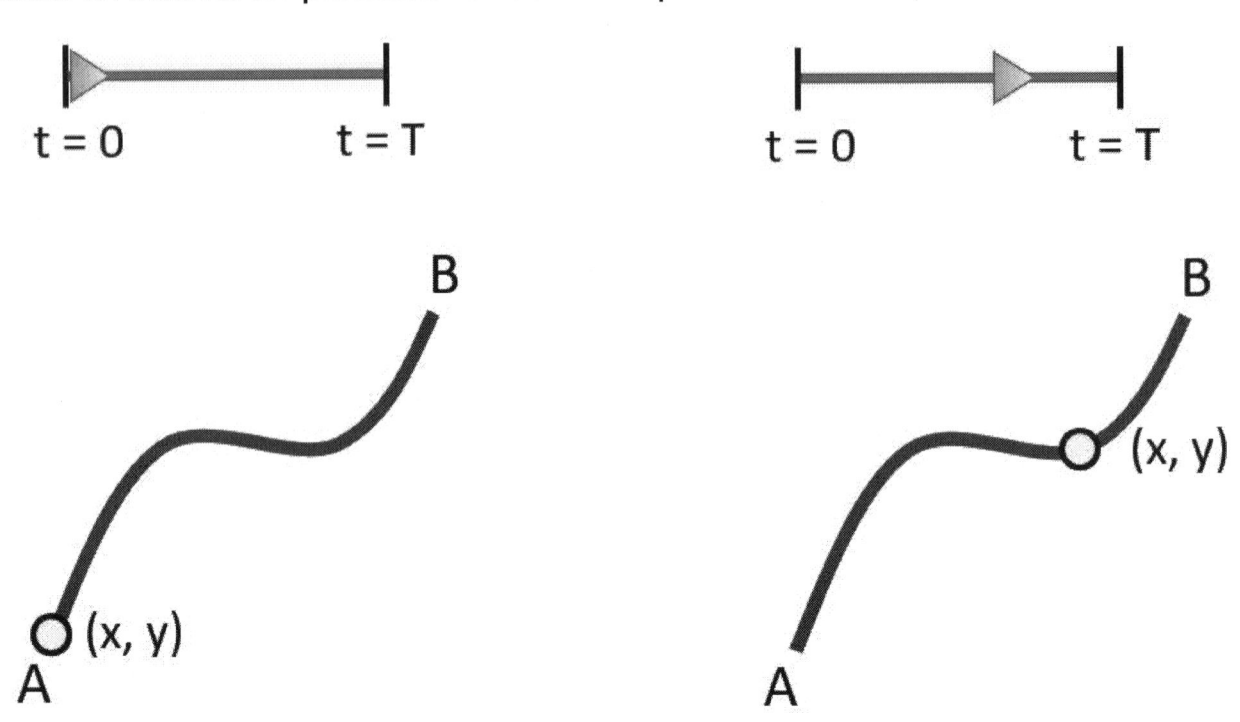

This is the basic idea behind parameterization of a curve. There is an added benefit to parameterization, it gives an orientation to the curve, which is important in many applications. There are umpteen no. of ways to parameterize a curve, but here are some standard forms.

Curve	Parametric Equations
Line segment from (x_0, y_0, z_0) to (x_1, y_1, z_0)	$x = (1-t)x_0 + tx_1$ $y = (1-t)y_0 + ty_1$, $0 \leq t \leq 1$ $z = (1-t)z_0 + tz_1$
Ellipse $\frac{x^2}{a^2} + \frac{y^2}{b^2} = 1$	$x = a \cos t$ $y = b \sin t$, $0 \leq t \leq 2\pi$
Circle $x^2 + y^2 = r^2$	$x = r \cos t$ $y = r \sin t$, $0 \leq t \leq 2\pi$
Function of the form $y = f(x)$	$x = t$ $Y = f(t)$

Going back to the line integral, we need a way to express **dl** in terms of **dt**, our new parameter.

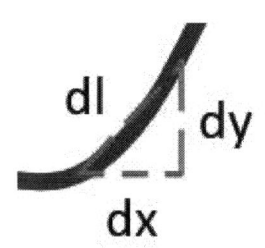

Consider a curve element **dl** as shown above. Using the Pythagoras theorem, we can express **dl** as,

$$dl = \sqrt{dx^2 + dy^2}$$

If the curve C is parameterized as **x=h(t), y=g(t)**, then the arc length **dl** is given by,

$$dl = \sqrt{(\frac{dx}{dt})^2 + (\frac{dy}{dt})^2}\, dt$$

And therefore the Line integral can be obtained as,

$$\int_a^b f(h(t), g(t)) \sqrt{(\frac{dx}{dt})^2 + (\frac{dy}{dt})^2}\, dt$$

The concept of parameterization and computing the line integral will become clear as we tackle some examples.

Solved Examples:

1. Compute the line integral of the function $f(x,y) = x^2 + y^2$ along the line segment from (0,0) to (5,12).

Line segment from (0,0) to (5,12) in the parameterized form is:

$$x = (1-t)0 + 5t = 5t$$
$$y = (1-t)0 + 12t = 12t$$

$$dl = \sqrt{\left(\frac{dx}{dt}\right)^2 + \left(\frac{dy}{dt}\right)^2}\, dt$$
$$= \sqrt{(5)^2 + (12)^2}\, dt$$
$$= 13\, dt$$

$$f(x,y) = x^2 + y^2 = 25t^2 + 144t^2$$
$$= 169t^2$$

$\therefore$ Line Integral $= \int_a^b f(h(t), g(t)) \sqrt{\left(\frac{dx}{dt}\right)^2 + \left(\frac{dy}{dt}\right)^2}\, dt$

$$= \int_0^1 169\, t^2 \times 13\, dt$$
$$= \frac{2197}{3}$$

2. Compute the line integral of the function f = 2xy along the portion of a unit circle that lies in the first quadrant.

The unit circle that lies in the first quadrant in the parameterized form is:

$$x = \cos t$$
$$y = \sin t$$
$$, 0 \leq t \leq \pi/2$$

$$dl = \sqrt{\left(\frac{dx}{dt}\right)^2 + \left(\frac{dy}{dt}\right)^2}\, dt$$
$$= \sqrt{(-\sin t)^2 + (\cos t)^2}\, dt$$
$$= dt$$

$$f(x,y) = 2xy = 2\sin t \cos t$$

$$\therefore \text{Line Integral} = \int_a^b f(h(t), g(t))\sqrt{\left(\frac{dx}{dt}\right)^2 + \left(\frac{dy}{dt}\right)^2}\, dt$$

$$= \int_0^{\pi/2} 2\sin t \cos t\, dt$$

Let $\sin t = u$, then $du = \cos t\, dt$ & $u \to 0$ to 1

$$= \int_0^1 2u\, du = 1$$

3. Compute the line integral of the function $f = x^2 + y^2$ along the line segment given by $x = -2 + 6t$ and $y = 5 + 8t$ for $0 \leq t \leq 1$.

In this problem, we don't have to parameterize the curve ourself, it is already given in the parameterized form.

$$x = -2 + 6t$$
$$y = 5 + 8t$$
$$, 0 \leq t \leq 1$$

$$dl = \sqrt{\left(\frac{dx}{dt}\right)^2 + \left(\frac{dy}{dt}\right)^2} \, dt$$
$$= \sqrt{(6)^2 + (8)^2} \, dt$$
$$= 10 \, dt$$

$$f(x,y) = x^2 + y^2 = 100t^2 + 56t + 29$$

∴ Line Integral $= \int_a^b f(h(t), g(t)) \sqrt{\left(\frac{dx}{dt}\right)^2 + \left(\frac{dy}{dt}\right)^2} \, dt$

$$= \int_0^1 (100t^2 + 56t + 29) 10 \, dt$$

$$= 10 \left[\frac{100}{3} t^3 + 28t^2 + 29t\right]_0^1$$

$$= \frac{2710}{3}$$

4. **Compute the line integral of the function f= 2x - 3y, first along the line segment from (-1,0) to (0,1), followed by the line segment from (0,1) to (0,3),**

In this problem, the curve is not a continuous one, so to compute the line integral we need to consider it as 2 separate curves and finally add the result.

Line segment from (-1,0) to (0,1) in the parameterized form is:

$$x = t-1$$
$$y = t$$

$$dl = \sqrt{\left(\frac{dx}{dt}\right)^2 + \left(\frac{dy}{dt}\right)^2}\, dt$$
$$= \sqrt{(1)^2 + (1)^2}\, dt$$
$$= \sqrt{2}\, dt$$

f(x,y) = 2x - 3y = -t -2

$$\therefore \text{Line Integral} = -\sqrt{2} \int_0^1 t+2\, dt$$
$$= \frac{-5\sqrt{2}}{2}$$

Line segment from (0,1) to (0,3) in the parameterized form is:

$$x = 0$$
$$y = 2t+1$$

$$dl = \sqrt{\left(\frac{dx}{dt}\right)^2 + \left(\frac{dy}{dt}\right)^2}\, dt$$
$$= \sqrt{(0)^2 + (2)^2}\, dt$$
$$= 2\, dt$$

$$f(x,y) = 2x - 3y = -6t - 3$$

$$\therefore \text{Line Integral} = -\int_0^1 (6t + 3)\, 2dt$$

$$= -12$$

Line Integral along the whole curve C

$$= -\frac{5\sqrt{2}}{2} - 12$$

B) Line Integral of a Vector field

Now let's move on to line integration of a vector field. The best way to interpret the line integral of a vector field is to consider it as the amount of work that a force field does on an object as it moves along a curve. When you try to move a block or a particle along a curve C in space at constant speed through a force field, a force always acts on the block, making it easier or harder to move the block depending on the directions of the forces at a certain point. If the force acts opposite to the direction of the path, then you have to do work to keep the block moving. On the other hand, if the direction of the force is in the direction of the path, then your job becomes easier because you're being aided by an outside force.

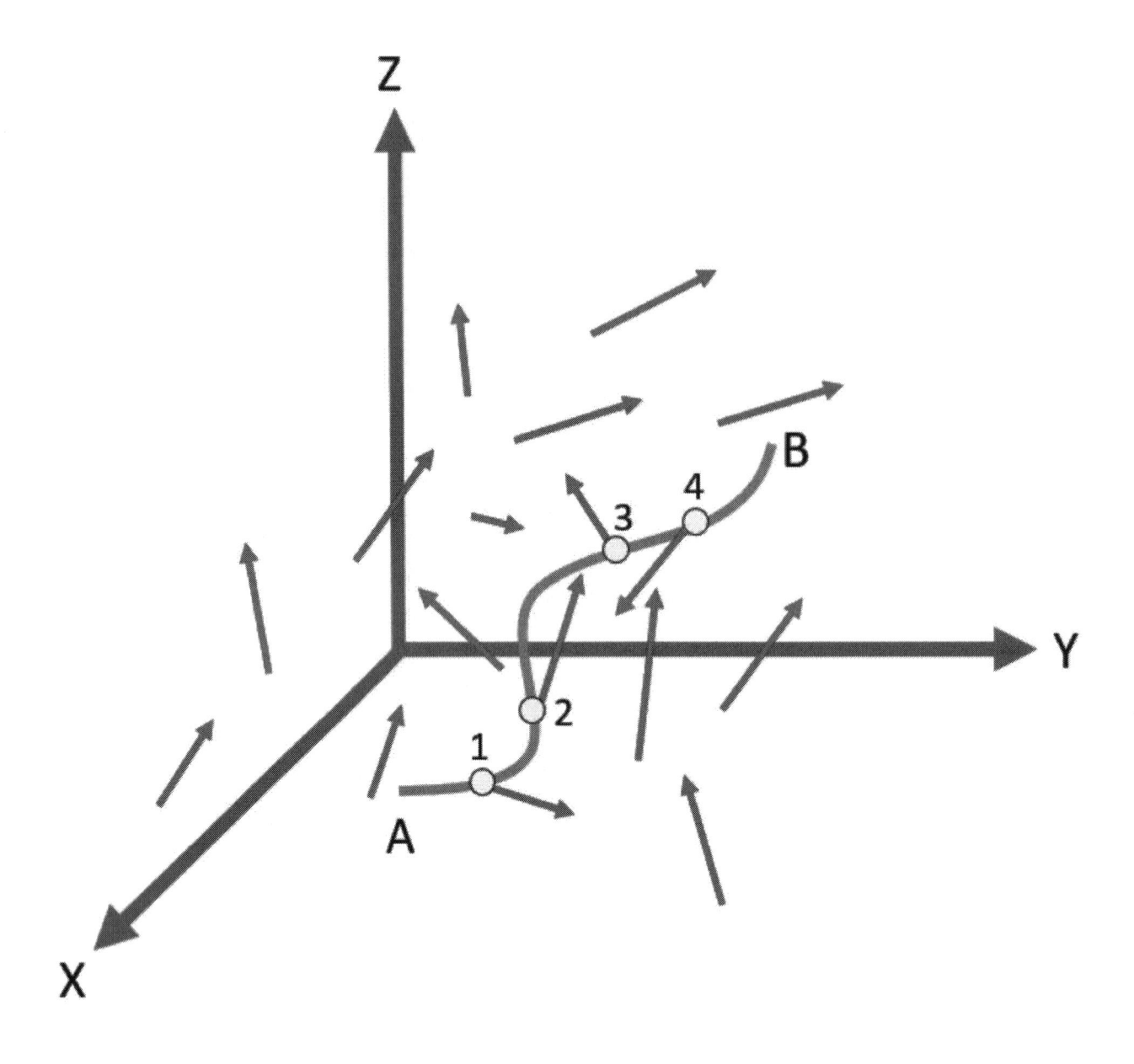

In the figure above, the line integral is to be evaluated along the curve AB. Now consider intermediate points (1,2,3,4) on the curve AB. At point 1, the force at the point is not completely aligned with the curve, but some component of it will be along the curve and hence it does some amount of work in moving the particle. Point 2 is also similar to point 1, here too some amount of work is done by the force. At point 3 however, the force at the point is perpendicular to the curve, which means it does not contribute to the motion of the particle and hence the work done by the force is zero. At point 4, the force at the point is acting opposite to the direction of the curve, so here the force makes it harder to move the particle along the curve i.e. negative work is being done. If we sum up the contributions of the force field at all such points along a path, what we get is the line integral.

Here unlike in the case of line integral of scalar fields, the direction of the curve is very important. In the above example, had we intended to move the particle along the path BA instead of AB, then the line integral will be the negative of line integral along AB.

So how do we calculate the line integral of a vector field mathematically? We do that by considering small segments along the curve as shown below. Let the unit tangent vector **dr** denote the direction of the curve at that segment. If vector **F** is the force acting on the particle, then **F.dr** denotes the work done by the force field on the particle at that segment. Remember how the dot product is a measure of how well 2 vectors work with each other.

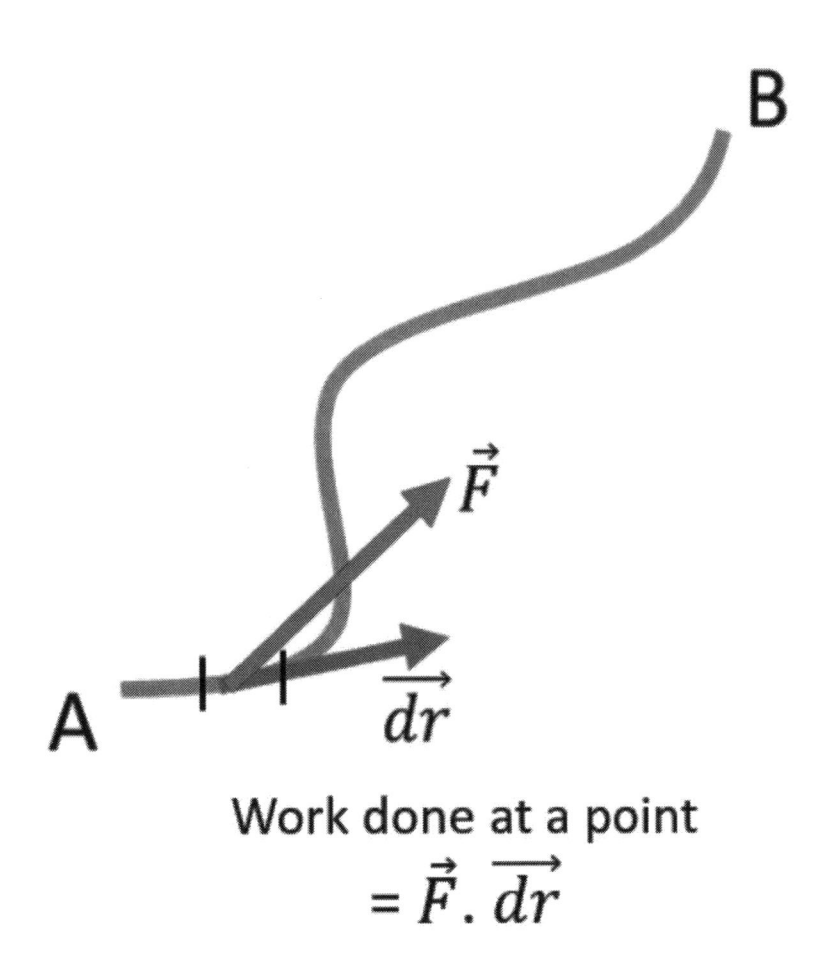

Work done at a point
$$= \vec{F} \cdot \vec{dr}$$

Now to evaluate the line integral all we need to do is to sum up the work done on the particle along all the segments along the curve. The integral takes care of that and hence the line integral is given by,

$$\text{Line integral} = \int_C \vec{F} \cdot \vec{dr}$$

Parameterization:

To compute the line integral of a vector field along a curve, we need to parameterize the curve as we did with scalar fields. Only difference being, that in this case the parameterized curve element has to be expressed as a vector.

$$\vec{r}(t) = x\hat{\imath} + y\hat{\jmath} + z\hat{k} \quad , a \leq t \leq b$$

And the Line integral can be obtained as,

$$\int_a^b \vec{F} \cdot \frac{d\vec{r}}{dt} \, dt$$

Solved Examples:

5. Evaluate the line integral of the vector field $F(x,y,z) = xz\,\hat{\imath} - yz\,\hat{k}$, along the line segment from $(-1,2,0)$ to $(3,0,1)$.

Line segment from (-1,2,0) to (3,0,1) in the parameterized form is:

$$x = (1-t)(-1) + 3t = 4t - 1$$
$$y = (1-t)2 + 0t = 2 - 2t \quad , 0 \leq t \leq 1$$
$$z = (1-t)0 + t = t$$

$$\vec{r}(t) = (4t-1)\hat{\imath} + (2-2t)\hat{\jmath} + t\hat{k}$$

$$\frac{d\vec{r}}{dt} = 4\hat{\imath} - 2\hat{\jmath} + \hat{k}$$

$$\vec{F}(t) = (4t-1)t\,\hat{\imath} - (2-2t)t\,\hat{k}$$
$$= (4t^2 - t)\hat{\imath} - (2t - 2t^2)\hat{k}$$

$$\therefore \text{Line Integral} = \int_a^b \vec{F} \cdot \frac{d\vec{r}}{dt}\, dt$$

$$= \int_0^1 ((4t^2 - t)\hat{\imath} - (2t - 2t^2)\hat{k}) \cdot (4\hat{\imath} - 2\hat{\jmath} + \hat{k})\, dt$$

$$= \int_0^1 (18t^2 - 6t)\, dt$$

$$= [6t^3 - 3t^2]_0^1$$

$$= 3$$

6. Compute the work that needs to be done in moving an object along the line segment from (1,2) to (0,0), through the force field $\vec{F} = y\hat{\imath} - x\hat{\jmath}$.

Line segment from (1,2) to (0,0) in the parameterized form is:

$$x = (1-t)(1) + 0t = 1 - t$$
$$y = (1-t)2 + 0t = 2 - 2t$$
$$, 0 \leq t \leq 1$$

$$\vec{r}(t) = (1-t)\hat{i} + (2-2t)\hat{j}$$

$$\frac{d\vec{r}}{dt} = -\hat{i} - 2\hat{j}$$

$$\vec{F}(t) = (2-2t)\hat{i} - (1-t)\hat{j}$$

$$\therefore \text{Line Integral} = \int_a^b \vec{F} \cdot \frac{d\vec{r}}{dt} \, dt$$

$$= \int_0^1 ((2-2t)\hat{i} - (1-t)\hat{j}) \cdot (-\hat{i} - 2\hat{j}) dt$$
$$= \int_0^1 (2t - 2 + 2 - 2t) dt$$
$$= 0$$

7. How much work is to be done to move an object in the vector field F=<y,3x> along the upper part of the ellipse $x^2/4+y^2=1$ from (2,0) to (-2,0)?

The top part of the ellipse in the parameterized form is:
$$x = 2\cos t$$
$$y = \sin t$$
$$, 0 \leq t \leq \pi$$

$$\vec{r}(t) = 2\cos t\,\hat{\imath} + \sin t\,\hat{\jmath}$$

$$\frac{d\vec{r}}{dt} = -2\sin t\,\hat{\imath} + \cos t\,\hat{\jmath}$$

$$\vec{F}(t) = \sin t\,\hat{\imath} + 6\cos t\,\hat{\jmath}$$

$$\therefore \text{Line Integral} = \int_a^b \vec{F} \cdot \frac{d\vec{r}}{dt}\,dt$$

$$= \int_0^\pi ((\sin t)(-2\sin t) + (6\cos t)(\cos t))\,dt$$
$$= \int_0^\pi (-2\sin^2 t + 6\cos^2 t)\,dt$$
$$= 2\pi$$

5.2 SURFACE INTEGRAL

A) Surface Integral of a Scalar field

For scalar functions, the surface integral is nothing but the volume enclosed by a surface and its projection on the function. The simplest case is when the area is planar or 2-dimensional as shown below.

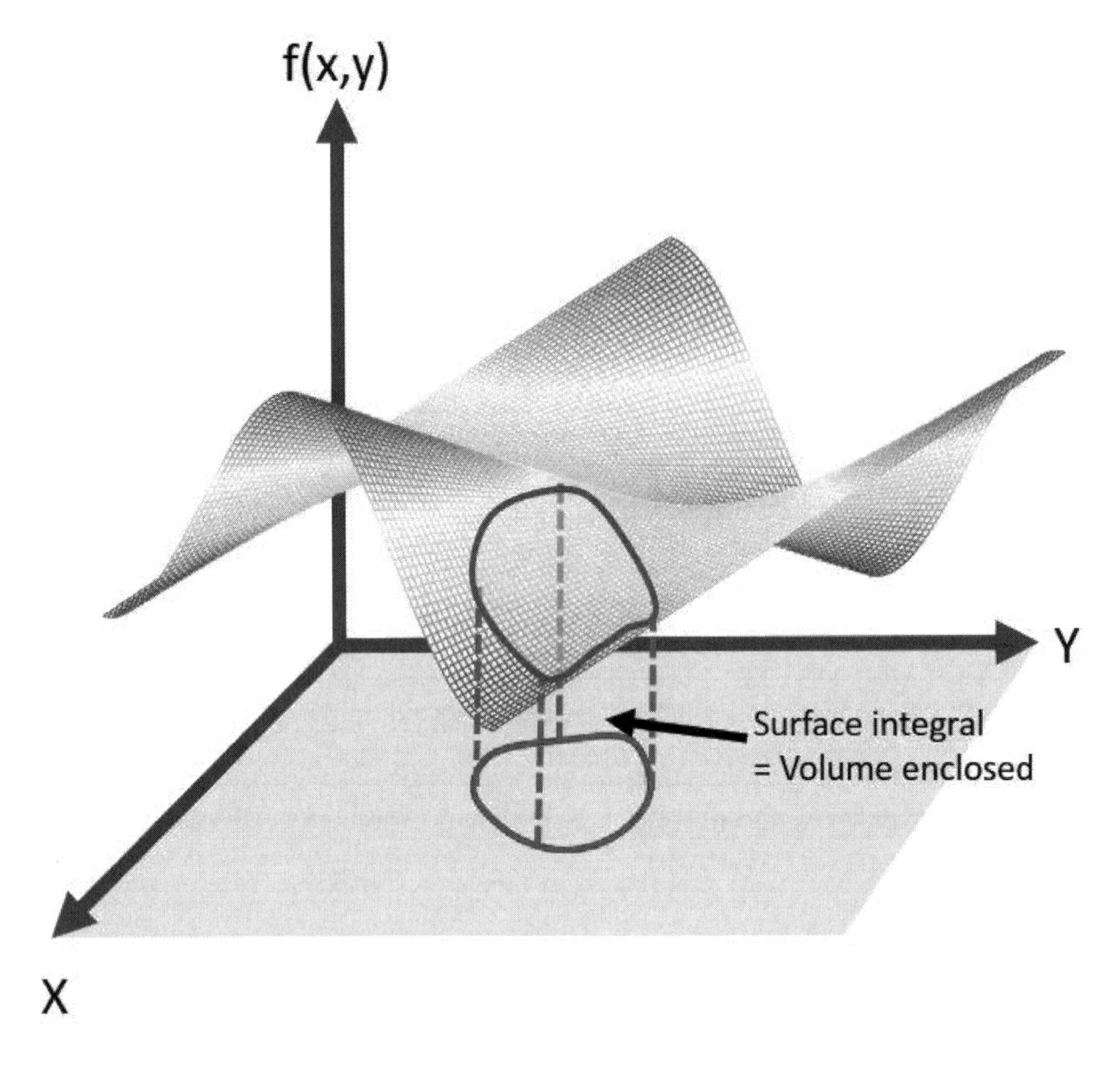

To find the surface integral in such a case, the surface is sliced using planes of small thickness either along the x-axis or the y-axis (along the x-axis in the figure). That way we can convert the surface into a bunch of curves or lines, allowing us to use the line integral.

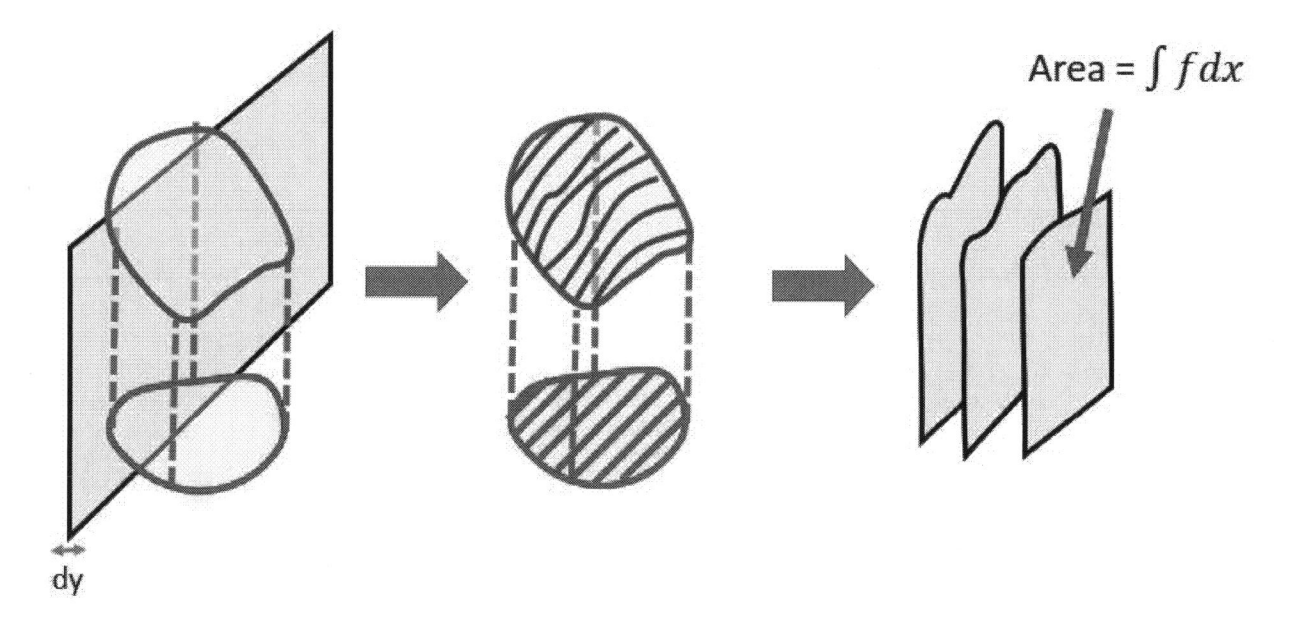

The sliced area is given by the line integral,

$$\int f \, dx$$

Here **dl** =**dx**, since the lines are straight and parallel to the x-axis. To obtain the volume, all we need to do is to combine these areas, which can be done by integrating them with respect to **dy**. Therefore, the surface integral is given by,

$$\iint f \, dx \, dy$$

This equation is called the double integral.

But not all surfaces are planar, some surfaces maybe spherical or conical or anything else, in such a case we would require a more generalized method to find the surface integral than the double integral. The idea here is to divide the whole surface into small differential surfaces each of area **dS**, then the volume corresponding to each differential area is the volume of the French fry shaped element (parallelopiped) above it.

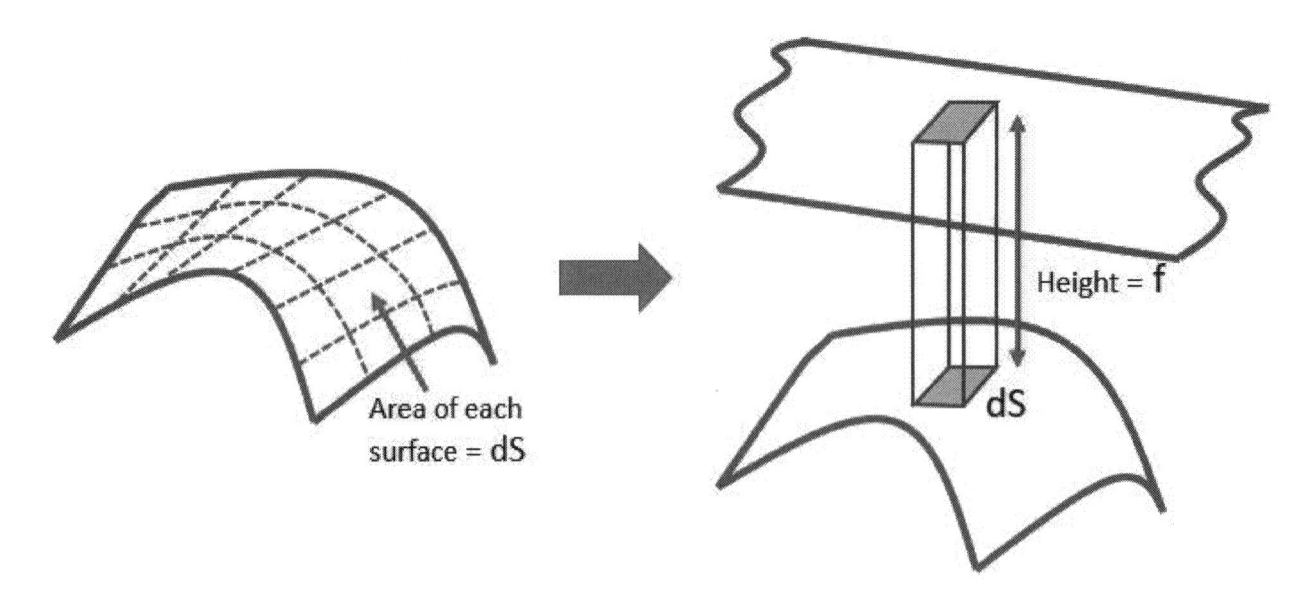

The volume of this element is **f dS** and hence the volume corresponding to the entire surface is simply the integral of this differential volume.

$$\iint_S f\, dS$$

Parameterization of the surface:

Just like with the line integral, to compute the surface integral we need to convert it into a convenient form. There are basically 2 methods to do this depending on the type of the surface.

If the surface is of the form **z = g(x,y)** (or can be expressed in such a form), then the surface integral can be obtained directly as,

$$\iint f(x, y, g(x,y)) \sqrt{\left(\frac{dg}{dx}\right)^2 + \left(\frac{dg}{dy}\right)^2 + 1}\ dx\, dy$$

This method can be used with any surface where a single variable (x, y or z) can be singled out in the equation.

For other type of surfaces, the standard double integral can't be used. In that case we need to parameterize the curve. For computing the line integral, we converted the whole equation into a function of a 3rd parameter **t**. Similarly for surface integrals, we need to express variables **x**, **y** and **z** as functions of 2 parameters **u** and **v**. Which essentially means we are mapping the 3d surface to a 2d plane.

First we need to parameterize the surface element in the form,

$$\vec{r}(u,v) = x(u,v)\,\hat{\imath} + y(u,v)\,\hat{\jmath} + z(u,v)\,\hat{k}$$

Then the line integral is given by,

$$\iint f(u,v) \left|\frac{d\vec{r}}{du} \times \frac{d\vec{r}}{dv}\right| du\, dv$$

This method is the general one and can be used for all type of surfaces. The limits of integration in these cases can be chosen according to the problem.

Solved Examples:

8. **Evaluate the surface integral of the function f = x+y+z over the portion of the plane x+2y+4z=4 lying in the first octant (x≥0, y≥0, z≥0).**

In this problem the variable **x** (or y or z as per convenience) can be singled out, therefore we can express the surface in the form **x = g(y,z)**.

$$x = g(y,z) = 4 - 2y - 4z$$

$$dS = \sqrt{\left(\frac{dg}{dy}\right)^2 + \left(\frac{dg}{dz}\right)^2 + 1}\ dy\ dz$$
$$= \sqrt{(-2)^2 + (-4)^2 + 1}\ dy\ dz$$
$$= \sqrt{21}\ dy\ dz$$

$$f(x, y, z) = x + y + z$$
$$\therefore f(g(y,z), y, z) = 4 - 2y - 4z + y + z$$
$$= 4 - y - 3z$$

Surface Integral $= \iint f(g(y,z), y, z)\ \sqrt{21}\ dy\ dz$

$$= \sqrt{21} \iint (4 - y - 3z)\ dy\ dz$$

Limits of Integration can be obtained as follows,

Surface: $x + 2y + 4z = 4$

Substituting $x = 0$, we get $2y + 4z = 4$
or $y + 2z = 2$

Therefore, $y = 2 - 2z$
Limits of y: $0 \to 2 - 2z$

Substituting $y = 0$ in $y = 2 - 2z$
Limits of z: $0 \to 2$

Here we are choosing to integrate with respect to **dy** first and then with respect to **dz**. Hence we chose the limits accordingly. Therefore the surface integral can be found as,

$$\text{Surface Integral} = \sqrt{21} \int_0^2 \int_0^{2-2z} (4 - y - 3z)\, dy\, dz$$

$$= \sqrt{21} \int_0^2 \left[4y - y^2 - 3yz\right]_0^{2-2z} dz$$

$$= \sqrt{21} \int_0^2 \left[6 - 10z + 4z^2\right] dz$$

$$= \sqrt{21} \left[6z - 10z^2 + \frac{4}{3}z^2\right]_0^2$$

$$= \frac{7\sqrt{21}}{3}$$

9. Evaluate the surface integral $\iint y \, dS$, over the portion of the cylinder $x^2+y^2 = 3$ that lies between z=0 and z=6.

The cone in the parameterized form is:

$$\vec{r}(z, \theta) = \sqrt{3} \cos\theta \, \hat{\imath} + \sqrt{3} \sin\theta \, \hat{\jmath} + z \, \hat{k}$$

$$0 \leq z \leq 6,$$
$$0 \leq \theta \leq 2\pi$$

$$\frac{d\vec{r}}{dz} = \hat{k}, \quad \frac{d\vec{r}}{d\theta} = -\sqrt{3} \sin\theta \, \hat{\imath} + \sqrt{3} \cos\theta \, \hat{\jmath}$$

$$\frac{d\vec{r}}{dz} \times \frac{d\vec{r}}{d\theta} = \begin{vmatrix} \hat{\imath} & \hat{\jmath} & \hat{k} \\ 0 & 0 & z \\ -\sqrt{3} & \sqrt{3} & 0 \end{vmatrix}$$

$$= -\sqrt{3} \cos\theta \, \hat{\imath} - \sqrt{3} \sin\theta \, \hat{\jmath}$$

$$\left| \frac{d\vec{r}}{dz} \times \frac{d\vec{r}}{d\theta} \right| = \sqrt{3}$$

Surface Integral $= \sqrt{3} \int_0^{2\pi} \int_0^6 \sqrt{3} \sin\theta \, dz \, d\theta$

$$= 3 \int_0^{2\pi} 6 \sin\theta \, d\theta$$

$$= [-18 \cos\theta]_0^{2\pi}$$

$$= 0$$

10.

Evaluate the surface integral of the function $f = \sqrt{1 + x^2 + y^2}$ over a surface S, parameterized by the position vector r(u,v)=u cosv i +u sinv j +v k, 0≤u≤2, 0≤v≤π.

$$\vec{r}(u,v) = u\cos v\ \hat{i} + u\sin v\ \hat{j} + v\ \hat{k}$$

$$\frac{d\vec{r}}{du} = \cos v\ \hat{i} + \sin v\ \hat{j},$$

$$\frac{d\vec{r}}{dv} = -u\sin v\ \hat{i} + u\cos v\ \hat{j} + \hat{k}$$

$$\frac{d\vec{r}}{du} \times \frac{d\vec{r}}{dv} = \begin{vmatrix} \hat{i} & \hat{j} & \hat{k} \\ \cos v & \sin v & 0 \\ -u\sin v & u\cos v & 1 \end{vmatrix}$$

$$= \sin v\ \hat{i} - \cos v\ \hat{j} + u\ \hat{k}$$

$$\left|\frac{d\vec{r}}{du} \times \frac{d\vec{r}}{dv}\right| = \sqrt{1+u^2}$$

$$F = \sqrt{1+x^2+y^2} = \sqrt{1+u^2}$$

$$\text{Surface Integral} = \int_0^\pi \int_0^2 1+u^2\ du\ dv$$

$$= \int_0^\pi \left[u + \frac{u^3}{3}\right]_0^2 dv$$

$$= \frac{14\pi}{3}$$

B) Surface Integral of a Vector field

The surface integral of a vector field represents the notion of Flux. The flux is a measure of the amount of field flowing (think of it as fluid flow) through a surface per unit time. Imagine you place a net in flowing water at some angle, then the surface integral measures the rate of flow through that net.

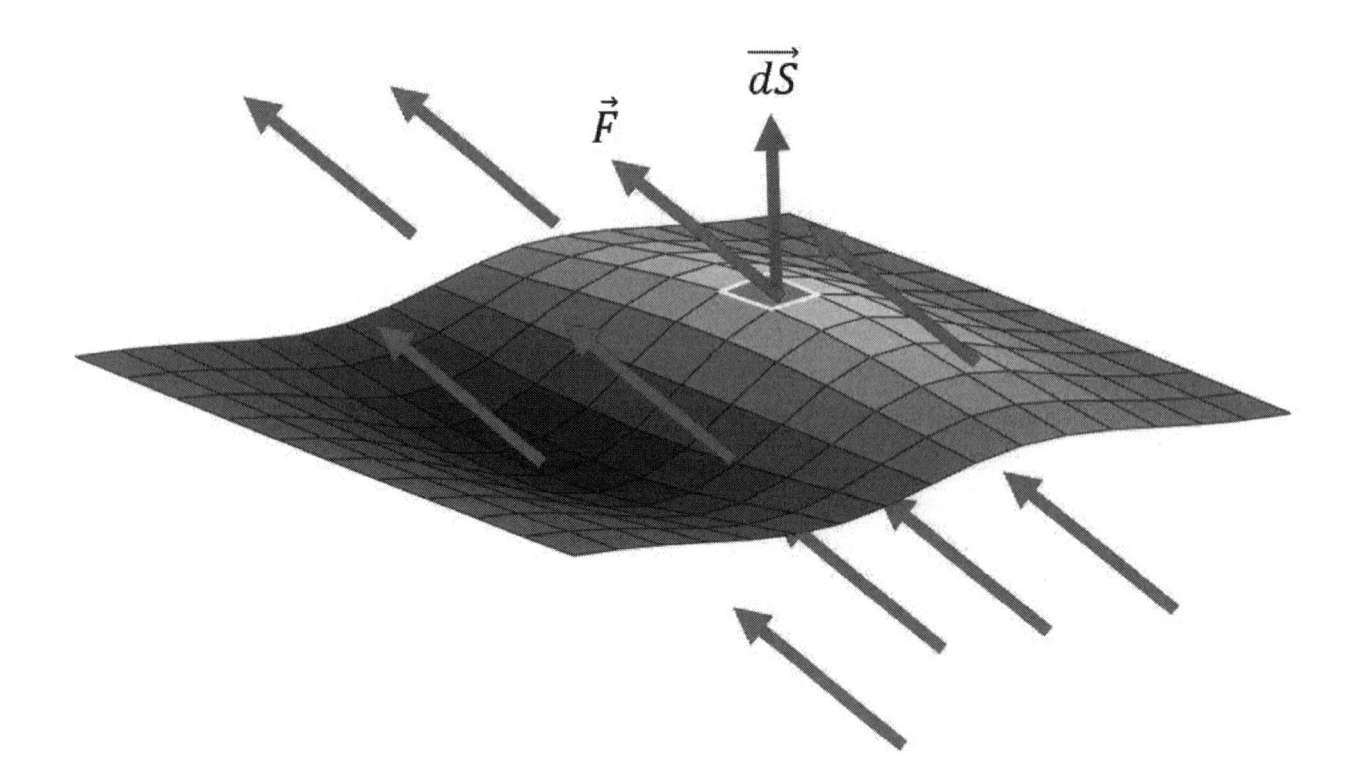

If the vector field is normal to the surface, the flow will be maximum and when the vector field is parallel to the surface there will be no flow. To calculate the total field flowing through the surface, consider a tiny area as shown in the figure above and its corresponding unit normal vector **dS** denotes the orientation of the tiny area. So the flux through the tiny area is given by **F.dS**, which is basically the perpendicular component of the vector field passing through the tiny area.

Therefore, the surface integral of a vector field (or the net flux) is given by,

$$\text{Surface integral} = \iint_S \vec{F} \cdot \vec{dS}$$

Parameterization:

Using the parameterized form, the Surface integral can be computed as,

$$\iint \vec{F} \cdot \left(\frac{\partial \vec{r}}{\partial u} \times \frac{\partial \vec{r}}{\partial v}\right) du\, dv$$

If the surface is of the form **z = g(x,y)** (or can be expressed in such a form), then the surface integral can be obtained directly as,

$$\iint \left(\vec{F} \cdot \left(-\frac{dg}{dx}\hat{i} - \frac{dg}{dy}\hat{j} + \hat{k}\right) dx\, dy\right)$$

Solved Examples:

11.

Evaluate the surface integral of the vector field F = -y i + x j over a surface S, parameterized by the position vector r(u,v)=u $\hat{i}$ +(v²-u) $\hat{j}$ +(u+v) $\hat{k}$, 0≤u≤3, 0≤v≤4.

$$\vec{r}(u, v) = u\,\hat{i} + (v^2 - u)\hat{j} + (u + v)\,\hat{k}$$

$$\frac{d\vec{r}}{du} = \hat{i} - \hat{j} + \hat{k},$$

$$\frac{d\vec{r}}{dv} = 2v\,\hat{j} + \hat{k}$$

$$\frac{d\vec{r}}{du} \times \frac{d\vec{r}}{dv} = \begin{vmatrix} \hat{i} & \hat{j} & \hat{k} \\ 1 & -1 & 1 \\ 0 & 2v & 1 \end{vmatrix}$$

$$= (-1-2v)\,\hat{i} - \hat{j} + 2v\,\hat{k}$$

$$\vec{F} = (u - v^2)\hat{i} + u\hat{j}$$

$$\vec{F} \cdot \left(\frac{d\vec{r}}{du} \times \frac{d\vec{r}}{dv}\right) = ((u - v^2)\hat{i} + u\hat{j}) \cdot ((-1-2v)\,\hat{i} - \hat{j} + 2v\,\hat{k})$$

$$= 2v^3 + v^2 - 2uv - 2u$$

Surface Integral $= \int_0^4 \int_0^3 2v^3 + v^2 - 2uv - 2u \, du \, dv$

$= \int_0^4 [2uv^3 + uv^2 - u^2v - u^2]_0^3 \, dv$

$= \int_0^4 [6v^3 + 3v^2 - 9v - 9] \, dv$

$= \left[\dfrac{3v^4}{2} + v^3 - \dfrac{9v^2}{2} - 9v\right]_0^4$

$= 340$

12.

Evaluate the surface integral of the vector field $F = 3x^2 \hat{i} - 2yx \hat{j} + 8 \hat{k}$ over the surface S that is the graph of $z = 2x - y$ over the rectangle [0, 2] × [0, 2].

The surface S in this problem is of the form **z=g(x,y)**, so we can calculate the surface integral directly without parameterization.

$$g(x,y) = 2x - y$$

$$\frac{dg}{dx} = 2, \frac{dg}{dy} = -1$$

$$-\frac{dg}{dx}\hat{i} - \frac{dg}{dy}\hat{j} + \hat{k} = -2\hat{i} + \hat{j} + \hat{k}$$

$$\vec{F} = 3x^2\hat{i} - 2yx\hat{j} + 8\hat{k}$$

$$\vec{F}\cdot(-\frac{dg}{dx}\hat{i} - \frac{dg}{dy}\hat{j} + \hat{k}) = (3x^2\hat{i} - 2yx\hat{j} + 8\hat{k})\cdot(-2\hat{i} + \hat{j} + \hat{k})$$

$$= -6x^2 - 2yx + 8$$

Surface Integral $= \int_0^2 \int_0^2 (-6x^2 - 2yx + 8)\, dx\, dy$

$$= \int_0^2 [-2x^3 - yx^2 + 8x]_0^2 \, dy$$

$$= \int_0^4 -4y\, dy$$

$$= [-2y^2]_0^2$$

$$= -8$$

13.

Evaluate the flux of the vector field F $=-y\hat{i}+x\hat{j}-z\hat{k}$ through the unit sphere $x^2+y^2+z^2=1$ that has downward orientation.

To parameterize the sphere surface, it is convenient to use spherical coordinates, which we introduce in the next chapter. For now just tag along.

The sphere in the parameterized form is:

$$\vec{r}(\phi, \theta) = \cos\phi \sin\theta \,\hat{\imath} + \sin\phi \sin\theta \,\hat{\jmath} + \cos\theta \,\hat{k}$$
$$0 \leq \phi \leq 2\pi,$$
$$0 \leq \theta \leq \pi$$

$$\frac{d\vec{r}}{d\phi} = -\sin\phi \sin\theta \,\hat{\imath} + \cos\phi \sin\theta \,\hat{\jmath},$$

$$\frac{d\vec{r}}{d\theta} = \cos\phi \cos\theta \,\hat{\imath} + \sin\phi \cos\theta \,\hat{\jmath} - \sin\theta \,\hat{k}$$

$$\frac{d\vec{r}}{d\phi} \times \frac{d\vec{r}}{d\theta} = \begin{vmatrix} \hat{\imath} & \hat{\jmath} & \hat{k} \\ -\sin\phi \sin\theta & \cos\phi \sin\theta & 0 \\ \cos\phi \cos\theta & \sin\phi \cos\theta & -\sin\theta \end{vmatrix}$$

$$= (-\cos\phi \sin^2\theta) \,\hat{\imath} + (-\sin\phi \sin^2\theta) \,\hat{\jmath}$$
$$+ (-\sin\phi \cos\theta) \,\hat{k}$$

$$\vec{F} = -sin\phi\ sin\theta\ \hat{i} + cos\phi\ sin\theta\ \hat{j} - cos\theta\ \hat{k}$$

$$\vec{F}\cdot\left(\frac{d\vec{r}}{d\phi} \times \frac{d\vec{r}}{d\theta}\right) = sin\theta\ cos^2\theta$$

Surface Integral $= \int_0^\pi \int_0^{2\pi} sin\theta\ cos^2\theta\ d\phi\ d\theta$

$$= \int_0^\pi [sin\theta\ cos^2\theta]_0^{2\pi} d\theta$$

$$= 2\pi \int_0^\pi -cos^2\theta\ d(cos\theta)$$

$$= -2\pi \left[\frac{cos^3\theta}{3}\right]_0^\pi$$

$$= \frac{4\pi}{3}$$

5.3 DIVERGENCE THEOREM

According to the Divergence Theorem *"The outward flux of a vector field through a closed surface is equal to the volume integral of the divergence of the vector field over the region enclosed by that closed surface"*.

Mathematically, the Divergence theorem can be expressed as:

$$\oint_S \vec{F}\cdot\overrightarrow{dS} = \int_V (\nabla\cdot\vec{F})\ dv$$

It is also known as Gauss's theorem or Ostrogradsky's theorem.

Although the statement and the mathematical formula may look complicated, the intuition behind this theorem is pretty straightforward.

Consider any 3 dimensional object placed in a vector field. We have gone with a potato shaped object as shown in the figure. Now let's consider the right hand side of the equation first. We have already learnt that the divergence of a vector field is measure of the outward going or the source like behavior at a point in the field. As this object under our consideration is placed in a vector field, every point (small volume) inside the object will have a divergence, either positive or negative or zero. In the figure below, we have shown the divergence at 3 points A, B and C. Now if we integrate (sum up) the divergence at all such points throughout the volume, what we get is the total divergence or the total source like nature of the object as a whole, which is essentially how much field originates from the object. And we know that the total flux originating from the object is equal the surface integral of the field across the full surface.

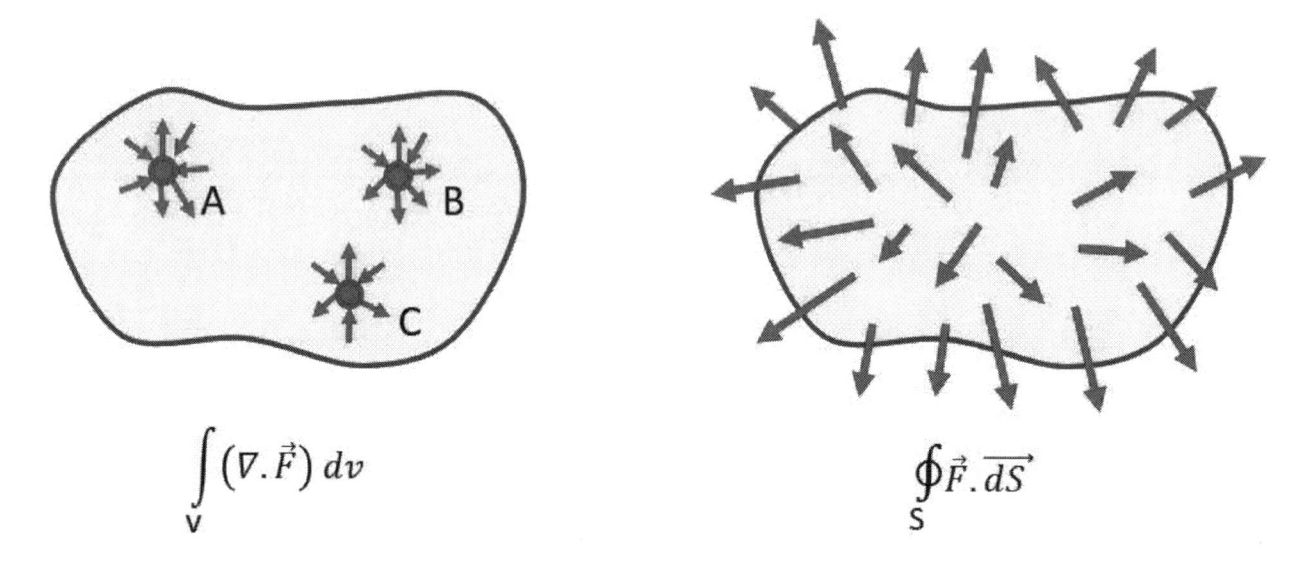

$$\int_V (\nabla \cdot \vec{F}) \, dv \qquad \oint_S \vec{F} \cdot \overrightarrow{dS}$$

The Divergence theorem in layman's terms can be stated as "*The total outward field flowing out from the surface of an object is the equal to the sum of the fields flowing out from every single point inside the volume of the object*".

What this theorem (and the one in the next section) really does is find a correlation between the smaller phenomenon that occurs on inside of an object to the larger phenomenon on the outer periphery. The divergence theorem provides a tool for converting surface integrals which are often difficult to compute into an easier volume integral. This is especially useful when we deal with some familiar shape or if the divergence results in a simple function.

An important thing to be noted is that the divergence theorem is only applicable to closed surfaces.

Solved Examples:

14. Using the Divergence theorem evaluate the surface integral of vector field $F = 3x\,\hat{i} + 2y\,\hat{j}$ over a surface S, which is a sphere $x^2+y^2+z^2 = 9$.

We could solve this problem by parameterizing as we did in the last section, but using the divergence theorem it's even simpler. First, by using the divergence theorem we can convert the surface integral into a volume integral.

$$\nabla \cdot \vec{F} = \frac{\partial(3x)}{\partial x} + \frac{\partial(2y)}{\partial y} + \frac{\partial(0)}{\partial z}$$
$$= 5$$

$$\iint_S (3x\,\hat{i} + 2y\,\hat{j}) \cdot \vec{dS} = \iiint_V 5\, dV$$
$$= 5 \times \text{Volume of the sphere}$$
$$= 180$$

15. Using the Divergence theorem evaluate the surface integral of vector field $F = \langle x,y,z \rangle$ over a surface S, which is the surface bounded by the cylinder $x^2+y^2=a^2$ and the planes $z = -1$ and $z = 1$.

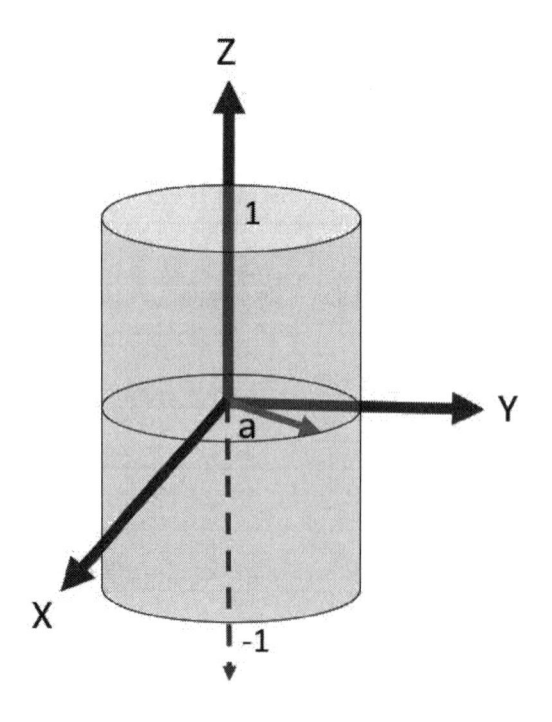

$$\nabla \cdot \vec{F} = \frac{\partial(x)}{\partial x} + \frac{\partial(y)}{\partial y} + \frac{\partial(z)}{\partial z}$$

$$= 3$$

$$\iint_S (x\hat{i} + y\hat{j} + z\hat{k}) \cdot \vec{dS} = \iiint_V 3\, dx\, dy\, dz$$

Using cylindrical coordinates,

$$= 3 \int_{-1}^{1} dz \int_{0}^{2\pi} d\theta \int_{0}^{a} r\, dr$$

$$= 6\pi a^2$$

In this problem we have used cylindrical coordinates, which is covered in the next chapter. Revisit this problem once you are familiar with cylindrical coordinates.

5.4 STOKE'S THEOREM

The Stoke's theorem states that *"the line integral of a vector field round that a closed path is equal to the surface integral of the curl of field over any surface bounded by that closed path"*.

Mathematically, the Stoke's theorem can be expressed as:

$$\oint_C \vec{F} \cdot \vec{dl} = \int_S (\nabla \times \vec{F}) \cdot \vec{dS}$$

The idea behind the Stoke's theorem is quite similar to the divergence theorem. Consider a planar surface placed in a vector field, there will be a rotational effect at every point on the surface which is given by the curl of the field. Now if we integrate (sum up) the curl at all points throughout the surface, what we get is the total rotational effect along the periphery due to the vector field. This is the same as the line integral of the vector field along the outer boundary.

The reason why this is true can be better understood with the help of the figure below. If you drop a ball near to the boundary close to point A, the ball will rotate in clockwise direction due to the curl at the point, given it's acting in isolation. But if you consider the curl at all points on the surface, a ball dropped near point A can no longer keep spinning at A, the curl at point B will inevitably push the ball from point A to point B and then it'll move from point B to point C and so on. In effect the ball dropped at the boundary will only keep moving along the boundary and never reach the inner surface. The curls on the inner surface cancel each other out, making it seem like the vector field is just acting along the periphery, not throughout the inner surface.

$$\int_S (\nabla \times \vec{F}) \cdot \vec{dS} \qquad \oint_C \vec{F} \cdot \vec{dl}$$

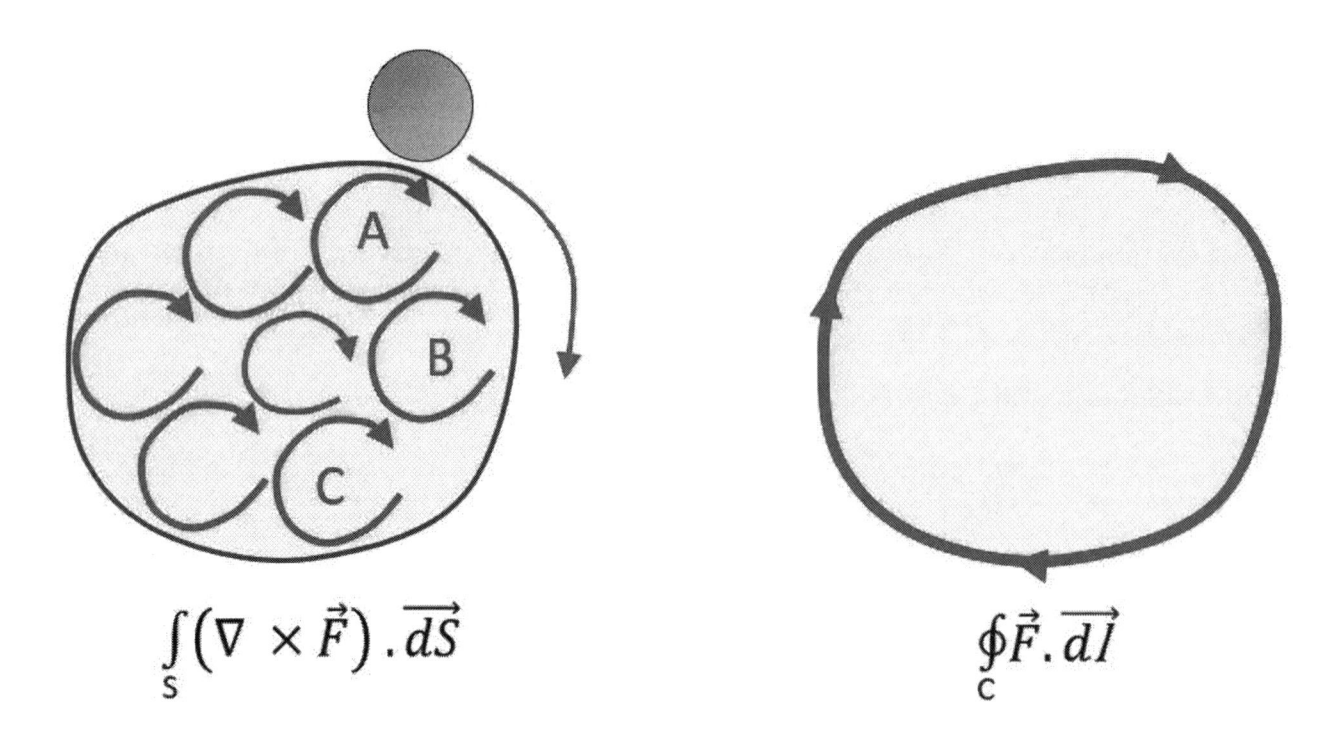

Solved Examples:

16. Show that the line integral of the vector field F = <yz, xz, xy> along any closed contour C is zero.

To prove this, we can convert the line integral to the surface using the divergence theorem.

$$\nabla \times \vec{F} = \begin{vmatrix} \hat{\imath} & \hat{\jmath} & \hat{k} \\ \dfrac{\partial}{\partial x} & \dfrac{\partial}{\partial y} & \dfrac{\partial}{\partial z} \\ yz & xz & xy \end{vmatrix}$$

$$= \left(\dfrac{\partial(xy)}{\partial y} - \dfrac{\partial(xz)}{\partial z}\right)\hat{\imath} + \left(\dfrac{\partial(yz)}{\partial z} - \dfrac{\partial(xy)}{\partial x}\right)\hat{\jmath} + \left(\dfrac{\partial(xz)}{\partial x} - \dfrac{\partial(yz)}{\partial y}\right)\hat{k}$$

$$= 0\hat{\imath} + 0\hat{\jmath} + 0\hat{k} = 0$$

$$\therefore \int_C \vec{F} \cdot \vec{dl} = \iint_S (\nabla \times \vec{F}) \cdot \vec{dS}$$

$$= 0$$

17. Find the surface integral of the curl of a vector field F = <2y cos z, e^x sinz, xe^y> over a surface S which is the hemisphere $x^2+y^2+z^2 = 9$ with z>0 oriented upwards.

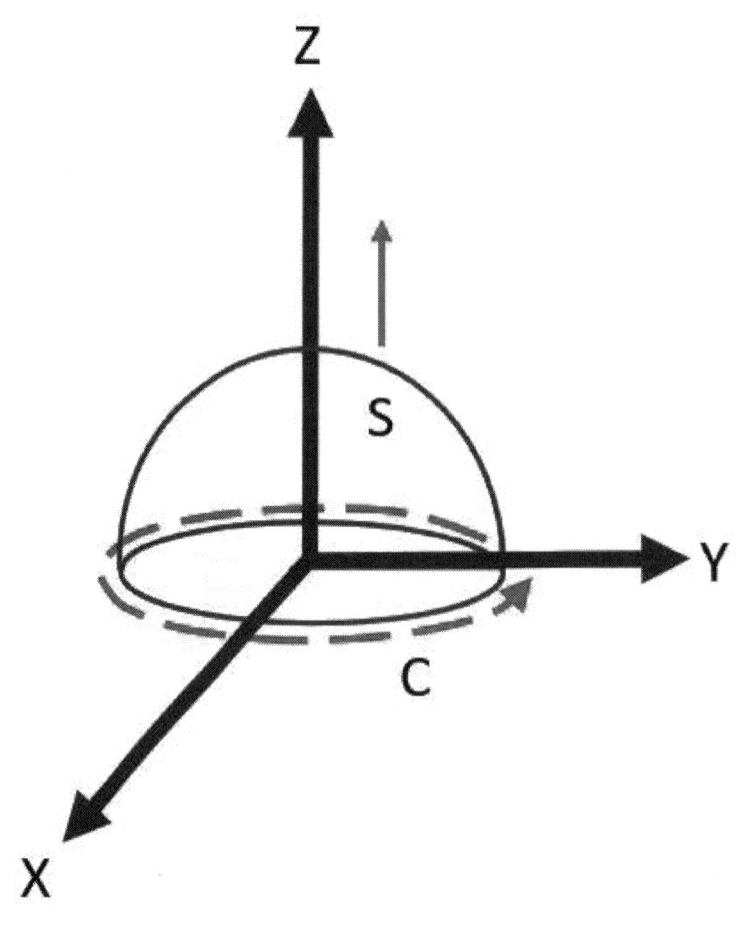

Using divergence theorem, we know that the line integral of the vector field around the closed path is equal to the surface integral of the curl of the field over the surface bound by the path. In this problem, the surface is a hemisphere as shown in the figure above and the closed path enclosing the surface is the perimeter of its circular base.

Substituting **z=0** in the equation of the hemisphere we can obtain the equation of the circular base.

Path C: $x^2 + y^2 = 9$

Circle in the parameterized form is:

$$\vec{r}(t) = 3\cos t\,\hat{i} + 3\sin t\,\hat{j}$$
$$0 \leq t \leq 2\pi$$

$$\frac{d\vec{r}}{dt} = -3\sin t\,\hat{i} + 3\cos t\,\hat{j}$$

$$\vec{F}(t) = 6\sin t\,\hat{i} + 0\hat{j} + (3\cos t)e^{\sin t}\,\hat{k}$$

$$\vec{F}(t) \cdot \frac{d\vec{r}}{dt} = -18\sin^2 t$$

Using Divergence Theorem,

$$\iint_S (\nabla \times \vec{F}) \cdot \vec{dS} = \int_C \vec{F} \cdot \vec{dl}$$

$$\int_C \vec{F} \cdot \vec{dl} = -18 \int_0^{2\pi} \sin^2 t\, dt$$

$$= -9 \int_0^{2\pi} 1 - \cos(2t)\, dt$$

$$= -9 \left[t - \frac{1}{2}\sin(2t)\right]_0^{2\pi}$$

$$= -18\pi$$

6. CURVILINEAR COORDINATE SYSTEMS

Up until now, we based all our definitions and explanations on the cartesian or the rectangular coordinate system. Although cartesian coordinates are very intuitive and easy to use, but when it comes to practical applications it is often more convenient to work with other coordinate systems. In this chapter we'll introduce Curvilinear coordinate systems, specifically the Cylindrical and the Spherical coordinate systems.

In any coordinate system a point can be defined by the intersection of 3 orthogonal surfaces. For curvilinear coordinate systems, a point is formed by the intersection of 3 curved planes.

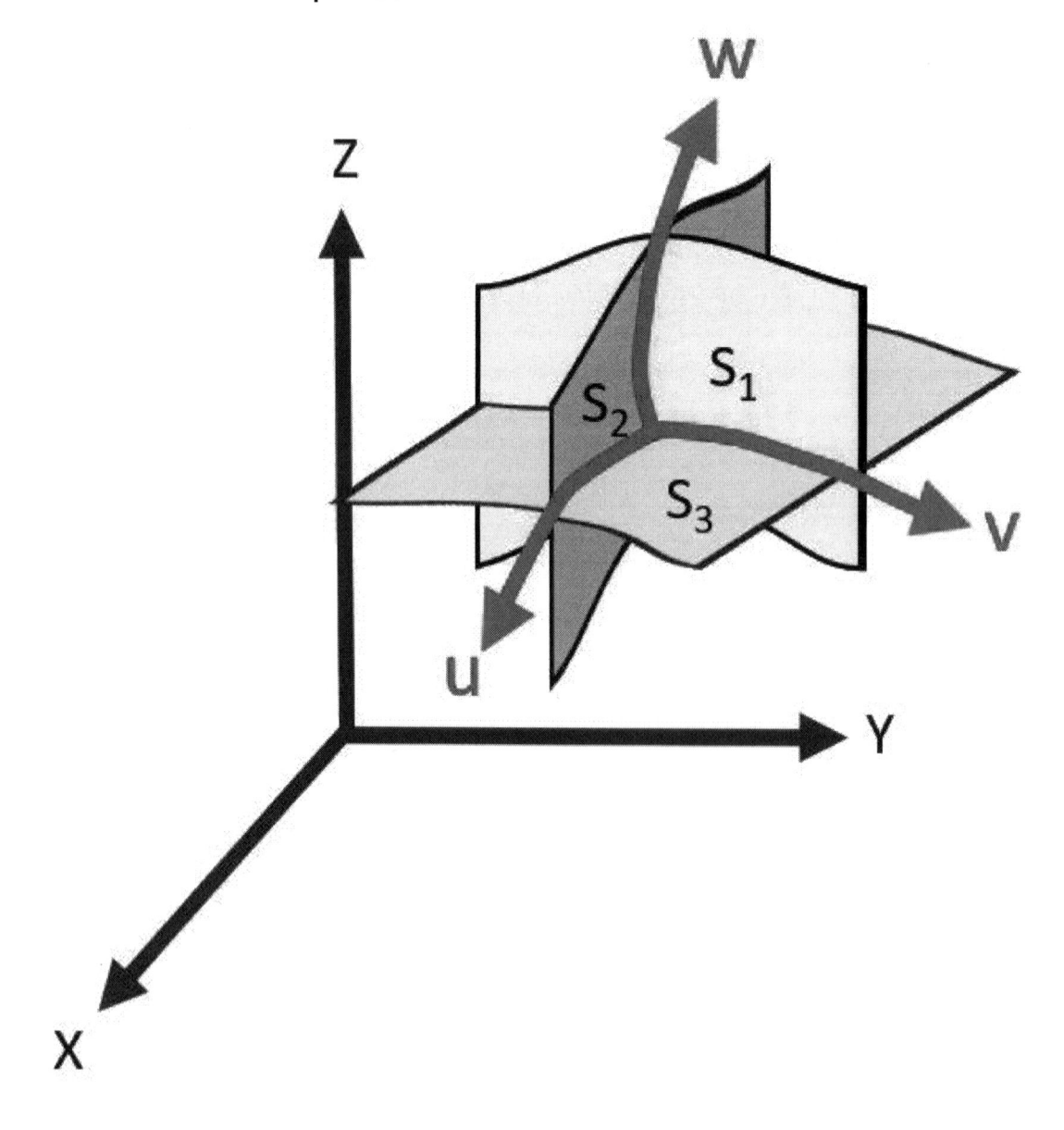

6.1 CYLINDRICAL COORDINATE SYSTEM

Cylindrical coordinate system is a special type of curvilinear coordinate system formed by the intersection of a cylinder, a plane and a half plane as shown below.

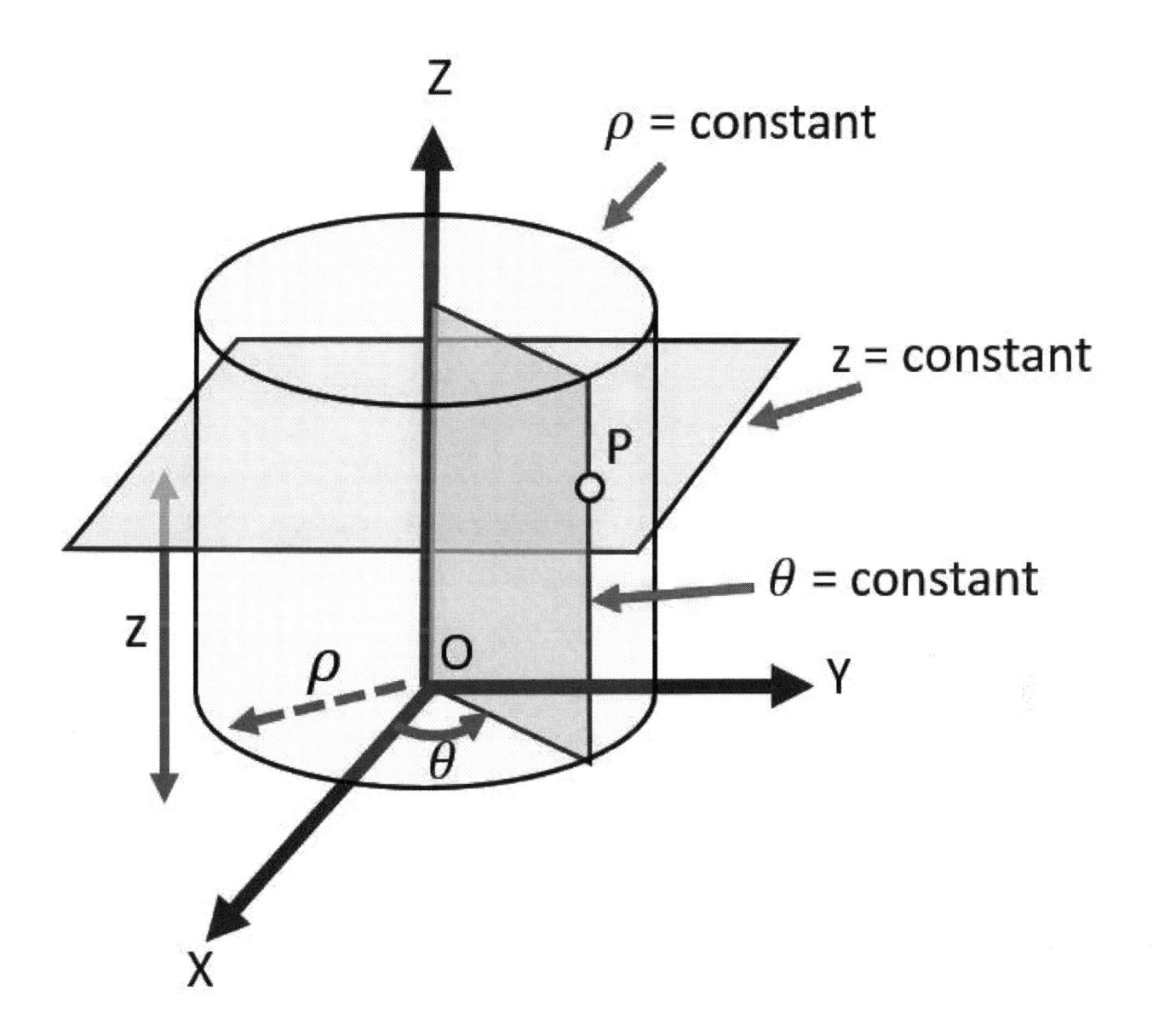

In this system, a point in space is specified in terms of,

- the distance of the point from the z-axis (ρ)
- the angle a half plane containing the point makes with the x- axis in the anticlockwise direction (θ). This angle is called the azimuth.

- the distance of the point from the xy plane (z) (same as in cartesian coordinates)

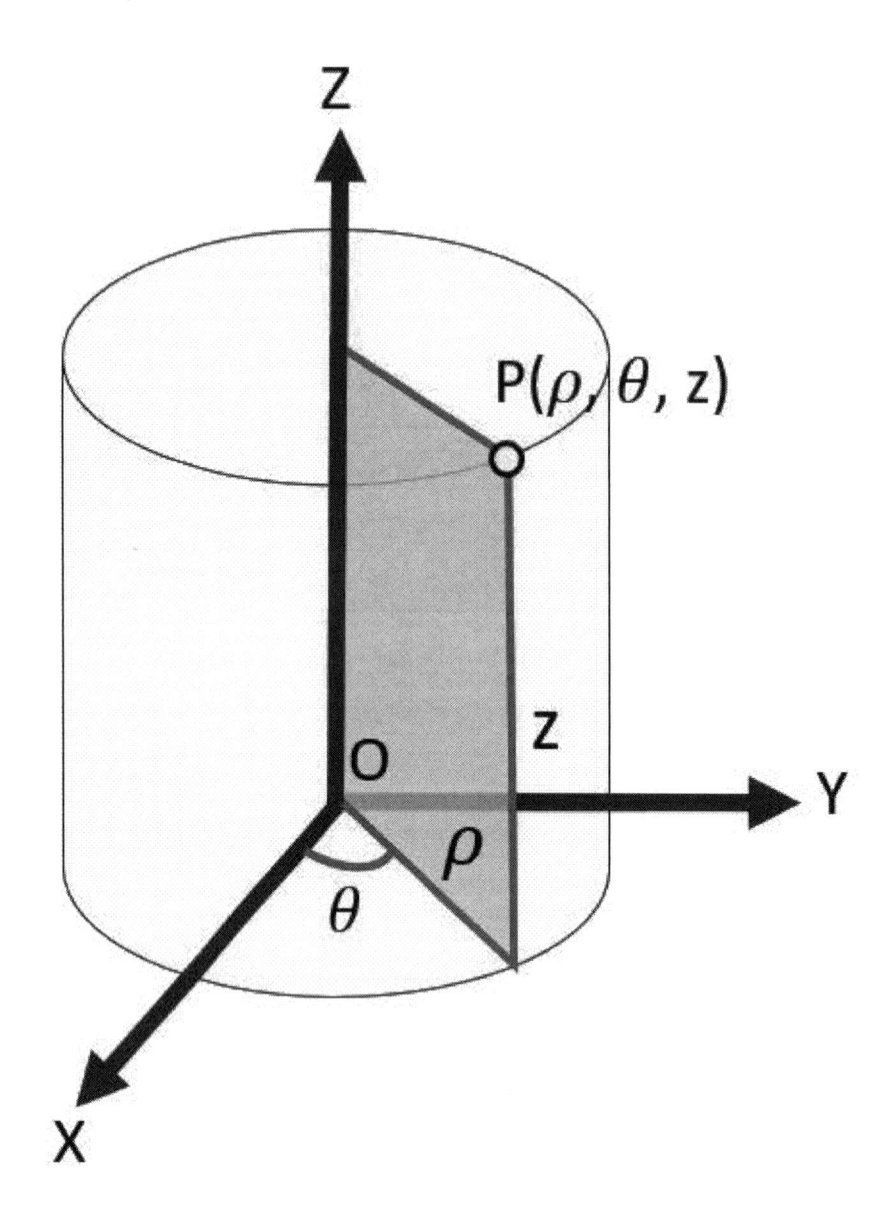

To cover the entire space, the radius of the cylinder ρ can be varied from 0 to ∞, the azimuth angle θ can be varied from 0 to 2π and the distance from the xy plane **z** can be varied from $-\infty$ to ∞.

Cylindrical to Cartesian coordinates:

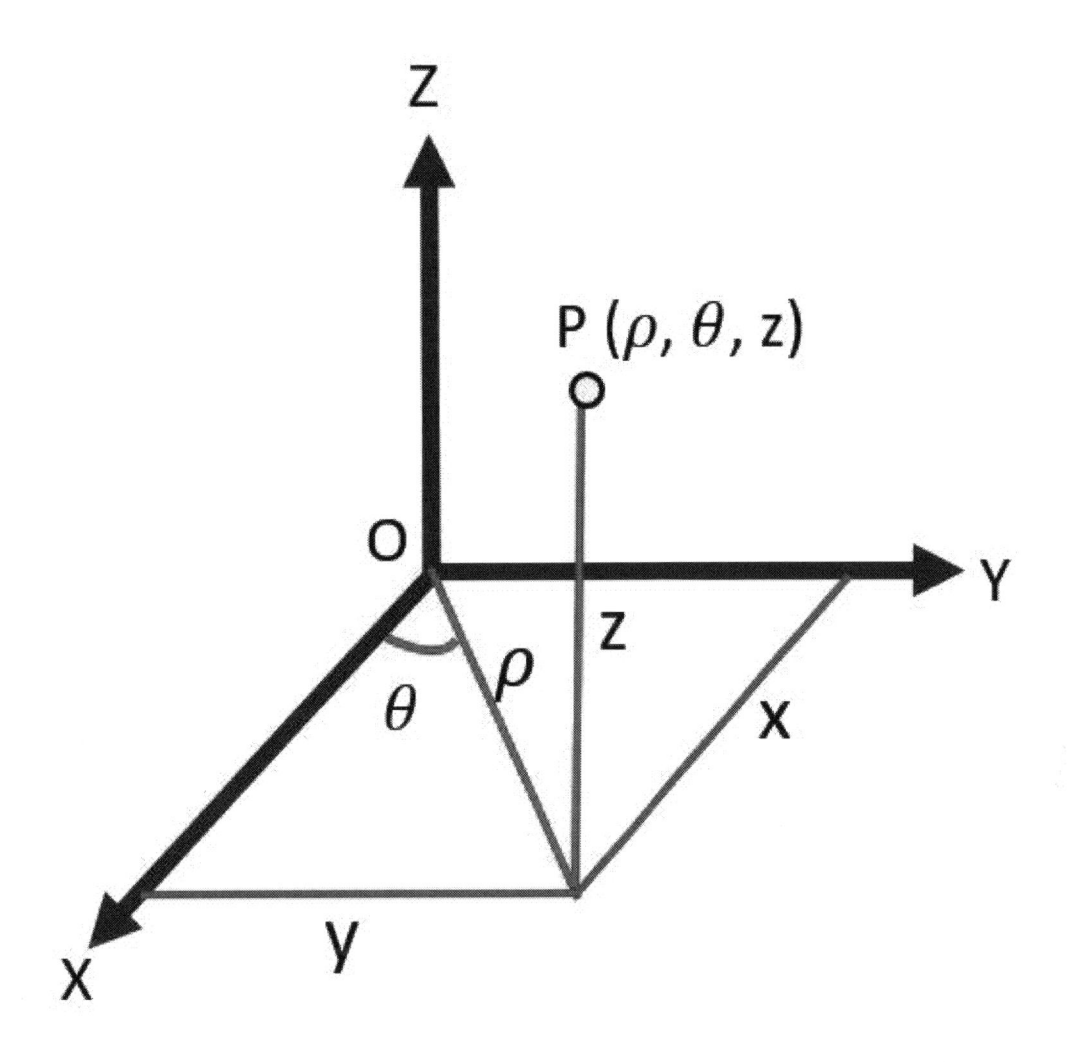

Cylindrical to Cartesian

$$x = \rho \cos\theta \qquad \rho = \sqrt{x^2 + y^2}$$
$$y = \rho \sin\theta \qquad \theta = \tan^{-1}(y/x)$$
$$z = z \qquad z = z$$

Unit Vectors in Cylindrical coordinates:

The concept of vectors remains unchanged irrespective of the coordinate system, but depending on the coordinate system used the 3 fundamental unit vectors change. In cylindrical coordinates, the fundamental unit vectors point in the directions of change of the three cylindrical coordinates ρ, θ, z as shown.

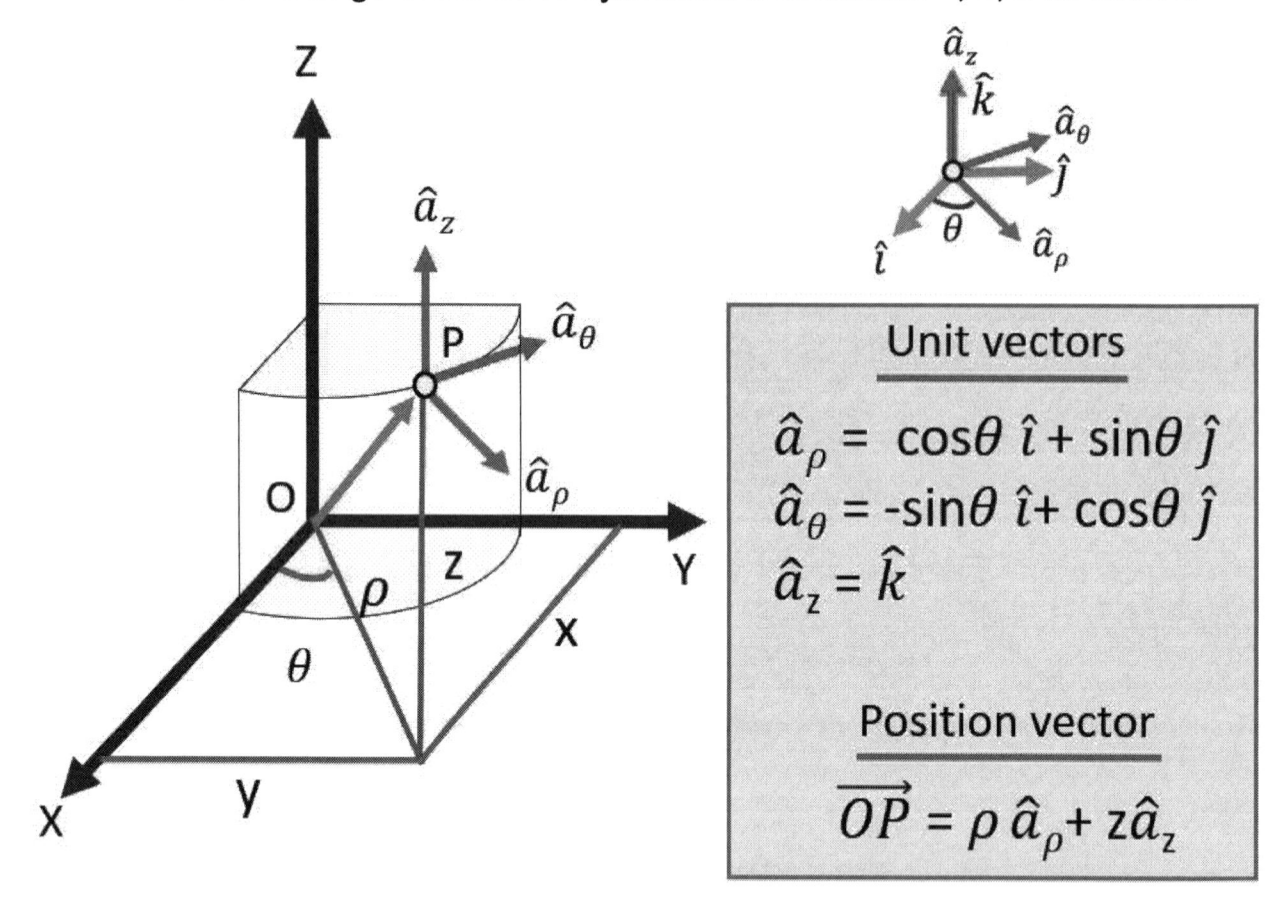

Unit vectors

$$\hat{a}_\rho = \cos\theta \, \hat{\imath} + \sin\theta \, \hat{\jmath}$$
$$\hat{a}_\theta = -\sin\theta \, \hat{\imath} + \cos\theta \, \hat{\jmath}$$
$$\hat{a}_z = \hat{k}$$

Position vector

$$\overrightarrow{OP} = \rho \, \hat{a}_\rho + z \hat{a}_z$$

Hence, in cylindrical coordinates any vector A can be expressed in terms of the fundamental unit vectors as,

$$\vec{A} = A_\rho \, \hat{a}_\rho + A_\theta \, \hat{a}_\theta + A_z \, \hat{a}_z$$

The one interesting thing about the fundamental unit vectors in cylindrical coordinates is that, unlike in cartesian coordinates, they are not universal. The unit vectors **a**$^\rho$ and **a**$^\theta$ are not constant, they have unit magnitude at all times but their directions depend on θ. Hence the derivatives of **a**$^\rho$ and **a**$^\theta$ with respect to θ are non zero.

$$\hat{a}_\rho = \cos\theta\, \hat{a}_x + \sin\theta\, \hat{a}_y$$

$$\frac{\partial(\hat{a}_\rho)}{\partial\theta} = -\sin\theta\, \hat{a}_x + \cos\theta\, \hat{a}_y$$

$$= \hat{a}_\theta$$

$$\hat{a}_\theta = -\sin\theta\, \hat{a}_x + \cos\theta\, \hat{a}_y$$

$$\frac{\partial(\hat{a}_\theta)}{\partial\theta} = -\cos\theta\, \hat{a}_x - \sin\theta\, \hat{a}_y$$

$$= -\hat{a}_\rho$$

We can summarize the derivatives of the unit vectors in cylindrical coordinates as,

$\frac{\partial(\hat{a}_\rho)}{\partial\rho} = 0$	$\frac{\partial(\hat{a}_\theta)}{\partial\rho} = 0$	$\frac{\partial(\hat{a}_z)}{\partial\rho} = 0$
$\frac{\partial(\hat{a}_\rho)}{\partial\theta} = \hat{a}_\theta$	$\frac{\partial(\hat{a}_\theta)}{\partial\theta} = -\hat{a}_\rho$	$\frac{\partial(\hat{a}_z)}{\partial\theta} = 0$
$\frac{\partial(\hat{a}_\rho)}{\partial z} = 0$	$\frac{\partial(\hat{a}_\theta)}{\partial z} = 0$	$\frac{\partial(\hat{a}_z)}{\partial z} = 0$

Differential Length, Volume in Cylindrical coordinates:

To find the differential length in cylindrical coordinates, we need to consider the differential change in any one variable, keeping the other variables unchanged.

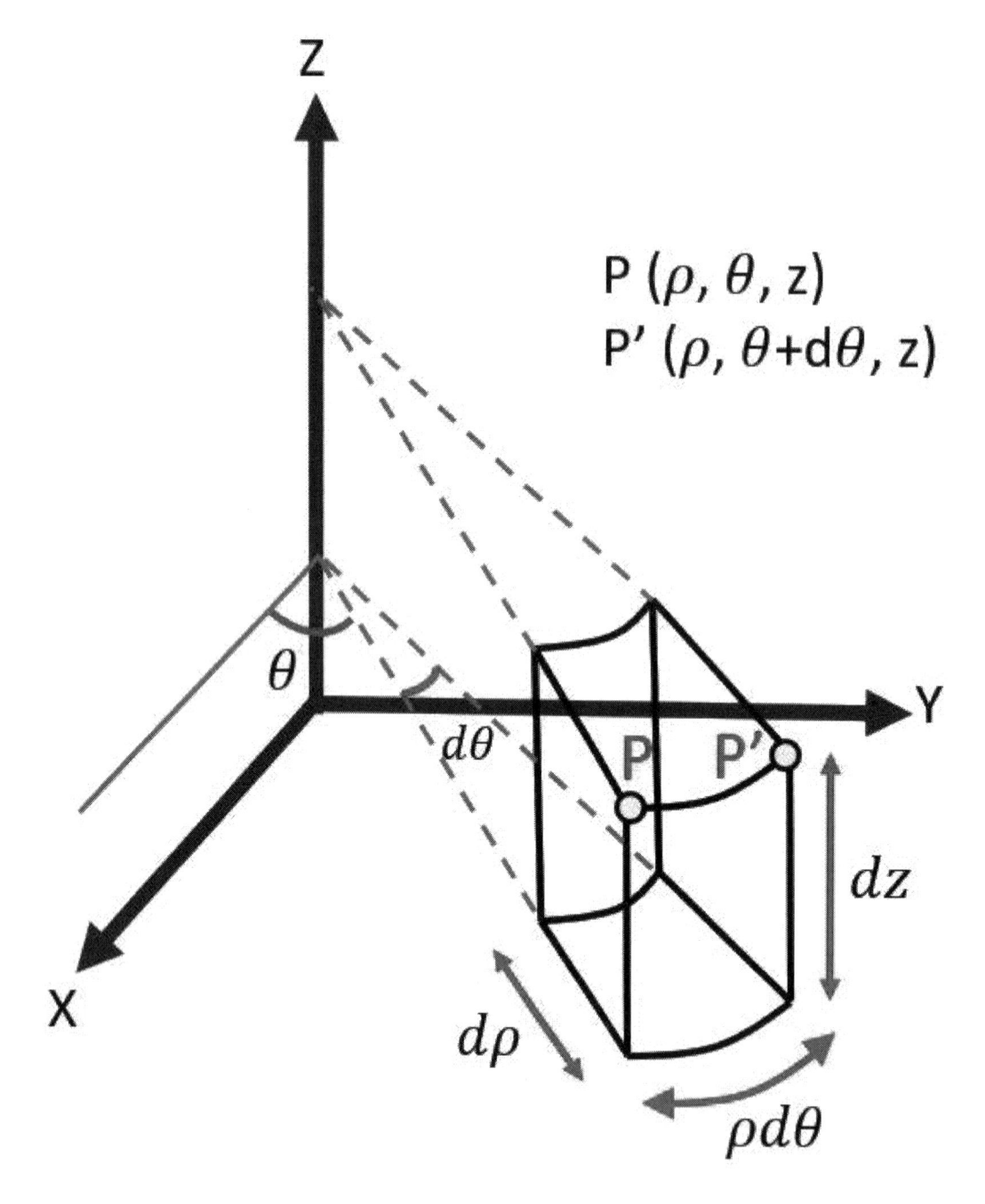

For instance, to find the differential change in length with respect to θ, assume that a point moves from P(ρ, θ, z) to P'(ρ, $\theta+d\theta$, z). Here the variables ρ and **z** variables remain constant, so there's just sideways movement. Hence the differential change in length is equal to the length of the arc PP' i.e. ***dl*** = $\rho d\theta$.

Similarly the differential change along ρ and **z** directions are $d\rho$ and dz respectively.

Therefore, in general the differential length vector ***dl*** is given by,

$$\vec{dl} = d\rho \, \hat{a}_\rho + \rho d\theta \, \hat{a}_\theta + dz \, \hat{a}_z$$

The volume of this differential element is given by,

$$dV = d\rho \, (\rho d\theta) dz = \rho \, d\rho \, d\theta \, dz$$

Del Operator in Cylindrical coordinates:

In cartesian coordinates the del operator is given by,

$$\nabla = \frac{\partial}{\partial x} \hat{i} + \frac{\partial}{\partial y} \hat{j} + \frac{\partial}{\partial z} \hat{k}$$

From this expression we can obtain the corresponding expression for cylindrical coordinates as follows. For that, first we need to express the derivative terms in the cylindrical coordinates using the chain rule.

Using Chain rule,
$$\frac{\partial}{\partial x} = \frac{\partial}{\partial \rho}\frac{\partial \rho}{\partial x} + \frac{\partial}{\partial \theta}\frac{\partial \theta}{\partial x}$$

$$\frac{\partial \rho}{\partial x} = \frac{\partial \sqrt{x^2+y^2}}{\partial x} = \frac{x}{\sqrt{x^2+y^2}} = \cos\theta$$

$$\frac{\partial \theta}{\partial x} = \frac{\partial \tan^{-1}(\frac{y}{x})}{\partial x} = \frac{-y}{x^2+y^2} = \frac{-\sin\theta}{\rho}$$

$$\therefore \frac{\partial}{\partial x} = (\cos\theta)\frac{\partial}{\partial \rho} + \left(\frac{-\sin\theta}{\rho}\right)\frac{\partial}{\partial \theta}$$

Similarly,
$$\frac{\partial}{\partial y} = (\sin\theta)\frac{\partial}{\partial \rho} + \left(\frac{\cos\theta}{\rho}\right)\frac{\partial}{\partial \theta}$$

In the above expressions, we have skipped the partial derivative term with respect to **z** because both ρ and θ are functions of **x** and **y** only and also the z coordinate is exactly the same in both systems. Now, we need to express the unit vectors in cartesian coordinates in terms of cylindrical coordinates.

$$\hat{\imath} = \cos\theta\ \hat{a}_\rho - \sin\theta\ \hat{a}_\theta$$
$$\hat{\jmath} = \sin\theta\ \hat{a}_\rho + \cos\theta\ \hat{a}_\theta$$
$$\hat{k} = \hat{a}_z$$

Plugging in all these values in the expression for the del operator in cartesian coordinates we get,

$$\nabla = (\cos\theta \, \hat{a}_\rho - \sin\theta \, \hat{a}_\theta)[(\cos\theta)\frac{\partial}{\partial \rho} + (\frac{-\sin\theta}{\rho})\frac{\partial}{\partial \theta}]$$
$$+ (\sin\theta \, \hat{a}_\rho + \cos\theta \, \hat{a}_\theta)[(\sin\theta)\frac{\partial}{\partial \rho} + (\frac{\cos\theta}{\rho})\frac{\partial}{\partial \theta}] + \frac{\partial}{\partial z}\hat{a}_z$$

Collecting all the common terms together, we get the final expression

$$\boxed{\nabla = \frac{\partial}{\partial \rho}\hat{a}_\rho + \frac{1}{\rho}\frac{\partial}{\partial \theta}\hat{a}_\theta + \frac{\partial}{\partial z}\hat{a}_z}$$

Gradient:

The expression for the gradient in cylindrical coordinates is,

$$\boxed{\nabla f = \frac{\partial f}{\partial \rho}\hat{a}_\rho + \frac{1}{\rho}\frac{\partial f}{\partial \theta}\hat{a}_\theta + \frac{\partial f}{\partial z}\hat{a}_z}$$

Where *f* is a scalar function expressed in cylindrical coordinates.

Divergence:

The divergence is the dot product of the del operator & a vector field.

$$\nabla \cdot \vec{A} = (\frac{\partial}{\partial \rho}\hat{a}_\rho + \frac{1}{\rho}\frac{\partial}{\partial \theta}\hat{a}_\theta + \frac{\partial}{\partial z}\hat{a}_z) \cdot (A_\rho \hat{a}_\rho + A_\theta \hat{a}_\theta + A_z \hat{a}_z)$$

Expanding this expression is not as simple as multiplying the respective components. That's because, as we mentioned earlier, the unit vectors in cylindrical coordinates are not universal and therefore the derivatives of these unit vectors also have to be taken into account.

Expanding the divergence expression using the product rule, we can obtain the divergence in cylindrical coordinates as,

$$\vec{\nabla}\cdot\vec{A} = \frac{1}{\rho}\frac{\partial(\rho A_\rho)}{\partial\rho} + \frac{1}{\rho}\frac{\partial(A_\theta)}{\partial\theta} + \frac{\partial(A_z)}{\partial z}$$

Curl:

The curl of a vector field in cylindrical coordinates can be obtained as,

$$\vec{\nabla}\times\vec{A} = \frac{1}{\rho}\begin{vmatrix} \hat{a}_\rho & \rho\hat{a}_\theta & \hat{a}_z \\ \frac{\partial}{\partial\rho} & \frac{\partial}{\partial\theta} & \frac{\partial}{\partial z} \\ A_\rho & \rho A_\theta & A_z \end{vmatrix}$$

Laplacian:

The Laplacian in cylindrical coordinates is given by,

$$\nabla^2 f = \frac{1}{\rho}\frac{\partial}{\partial\rho}\left(\rho\frac{\partial f}{\partial\rho}\right) + \frac{1}{\rho^2}\frac{\partial^2 f}{\partial\theta^2} + \frac{\partial^2 f}{\partial z^2}$$

There are rigorous mathematical techniques to derive these expressions, but these results itself are sufficient as far as application is concerned.

Solved Examples:

1. **Convert the point P(1,1,3) from cartesian to cylindrical coordinates.**

$$\rho = \sqrt{x^2 + y^2} = \sqrt{1^2 + 1^2} = \sqrt{2}$$
$$\theta = \tan^{-1}(y/x) = \tan^{-1}(1/1) = 45°$$
$$z = 3$$

Therefore in cylindrical coordinates the point is P(1.41, 45°, 3)

2. Convert the point P(5, 30°, 4) from cylindrical to cartesian coordinates.

$$x = \rho \cos\theta = 5 \cos 30 = 4.33$$
$$y = \rho \sin\theta = 5 \sin 30 = 2.5$$
$$z = z = 4$$

Therefore in cartesian coordinates the point is P(4.33, 2.5, 4).

3. A vector field is given as A = (x/r) $\hat{i}$ + (y/r) $\hat{j}$ + (z/r) $\hat{k}$, where r = $(x^2 + y^2 + z^2)^{1/2}$. Convert the vector field to cylindrical coordinates.

$$\vec{A} = \frac{x}{r}\hat{i} + \frac{y}{r}\hat{j} + \frac{z}{r}\hat{k},$$

where $r = \sqrt{x^2 + y^2 + z^2}$

Expressing x, y, z in cylindrical coordinates

$$x = \rho \cos\theta$$
$$y = \rho \sin\theta$$
$$z = z$$

Expressing unit vectors in cylindrical coordinates

$$\hat{i} = \cos\theta\, \hat{a}_\rho - \sin\theta\, \hat{a}_\theta$$
$$\hat{j} = \sin\theta\, \hat{a}_\rho + \cos\theta\, \hat{a}_\theta$$
$$\hat{k} = \hat{a}_z$$

$$r = \sqrt{x^2 + y^2 + z^2}$$
$$= \sqrt{\rho^2 \cos^2\theta + \rho^2 \sin^2\theta + z^2}$$
$$= \sqrt{\rho^2 + z^2}$$

$$\therefore \vec{A} = \frac{x}{r}\hat{i} + \frac{y}{r}\hat{j} + \frac{z}{r}\hat{k}$$

$$= \frac{1}{r}[x(\cos\theta\, \hat{a}_\rho - \sin\theta\, \hat{a}_\theta) + y(\sin\theta\, \hat{a}_\rho + \cos\theta\, \hat{a}_\theta) + z\hat{a}_z]$$

$$= \frac{1}{r}[(x\cos\theta + y\sin\theta)\hat{a}_\rho + (y\cos\theta - x\sin\theta)\hat{a}_\theta + z\hat{a}_z]$$

$$= \frac{\rho\hat{a}_\rho + z\hat{a}_z}{r}$$

$$= \frac{\rho\hat{a}_\rho + z\hat{a}_z}{\sqrt{\rho^2 + z^2}}$$

4. Find the gradient of the scalar function $f = \rho^2 z \cos 2\theta$.

$$f = \rho^2 z \cos2\theta$$

$$\nabla f = \frac{\partial f}{\partial \rho} \hat{a}_r + \frac{1}{\rho} \frac{\partial f}{\partial \theta} \hat{a}_\theta + \frac{\partial f}{\partial z} \hat{a}_z$$

$$\frac{\partial f}{\partial \rho} = 2\rho z \cos2\theta$$

$$\frac{\partial f}{\partial \theta} = -2\rho^2 z \sin2\theta$$

$$\frac{\partial f}{\partial z} = \rho^2 \cos2\theta$$

$$\nabla f = 2\rho z \cos2\theta \, \hat{a}_\rho - 2\rho^2 z \sin2\theta \, \hat{a}_\theta + \rho^2 \cos2\theta \, \hat{a}_z$$

5. Obtain the divergence of the vector field F = <$\rho\sin\theta$, $\rho^2 z$, $z\cos\theta$>.

$$\vec{F} = \rho \sin\theta \, \hat{a}_\rho + \rho^2 z \, \hat{a}_\theta + z \cos\theta \, \hat{a}_z$$

$$\nabla \cdot \vec{F} = \frac{1}{\rho} \frac{\partial(\rho F_\rho)}{\partial \rho} + \frac{1}{\rho} \frac{\partial(F_\theta)}{\partial \theta} + \frac{\partial(F_z)}{\partial z}$$

$$= \frac{1}{\rho} \frac{\partial(\rho^2 \sin\theta)}{\partial \rho} + \frac{1}{\rho} \frac{\partial(\rho^2 z)}{\partial \theta} + \frac{\partial(z \cos\theta)}{\partial z}$$

$$= 2\sin\theta + \cos\theta$$

6. Obtain the curl of the vector field F = <$\rho\sin\theta$, $\rho^2 z$, $z\cos\theta$>.

$$\vec{F} = \rho \sin\theta \, \hat{a}_\rho + \rho^2 z \, \hat{a}_\theta + z \cos\theta \, \hat{a}_z$$

$$\nabla \times \vec{F} = \frac{1}{\rho} \begin{vmatrix} \hat{a}_\rho & \rho\hat{a}_\theta & \hat{a}_z \\ \frac{\partial}{\partial \rho} & \frac{\partial}{\partial \theta} & \frac{\partial}{\partial z} \\ F_\rho & \rho F_\theta & F_z \end{vmatrix}$$

$$= \frac{1}{\rho} [(\frac{\partial(F_z)}{\partial \theta} - \frac{\partial(\rho F_\theta)}{\partial z})\hat{a}_\rho +$$

$$(\frac{\partial(F_\rho)}{\partial z} - \frac{\partial(F_z)}{\partial \rho})\rho\hat{a}_\theta + (\frac{\partial(\rho F_\theta)}{\partial r} - \frac{\partial(F_\rho)}{\partial \theta})\hat{a}_z]$$

$$= \frac{1}{\rho} [(\frac{\partial(z\cos\theta)}{\partial \theta} - \frac{\partial(\rho^3 z)}{\partial z})\hat{a}_\rho +$$

$$(\frac{\partial(\rho \sin\theta)}{\partial z} - \frac{\partial(z \cos\theta)}{\partial \rho})\rho\hat{a}_\theta + (\frac{\partial(\rho^3 z)}{\partial \rho} - \frac{\partial(\rho \sin\theta)}{\partial \theta})\hat{a}_z]$$

$$= \frac{1}{\rho} [(-z \sin\theta - \rho^3)\hat{a}_\rho + (3\rho^2 z - \rho \cos\theta)\hat{a}_z]$$

$$= -\frac{1}{\rho}(z \sin\theta + \rho^3)\hat{a}_\rho + (3\rho z - \cos\theta)\hat{a}_z$$

7. Find the gradient of the scalar function $f = \rho z \cos\theta$.

$$f = \rho z \cos\theta$$

$$\nabla^2 f = \frac{1}{\rho}\frac{\partial}{\partial \rho}\left(\rho \frac{\partial f}{\partial \rho}\right) + \frac{1}{\rho^2}\frac{\partial^2 f}{\partial \theta^2} + \frac{\partial^2 f}{\partial z^2}$$

$$\rho \frac{\partial f}{\partial \rho} = \rho z \cos\theta$$

$$\frac{1}{\rho}\frac{\partial}{\partial \rho}\left(\rho \frac{\partial f}{\partial \rho}\right) = \frac{z \cos\theta}{\rho}$$

$$\frac{1}{\rho^2}\frac{\partial^2 f}{\partial \theta^2} = \frac{-z \cos\theta}{\rho}$$

$$\frac{\partial^2 f}{\partial z^2} = 0$$

$$\therefore \nabla^2 f = \rho z \cos\theta$$

6.2 SPHERICAL COORDINATE SYSTEM

Cylindrical coordinate system is another type of curvilinear coordinate system, which is formed by the intersection of a sphere, a cone and a half plane.

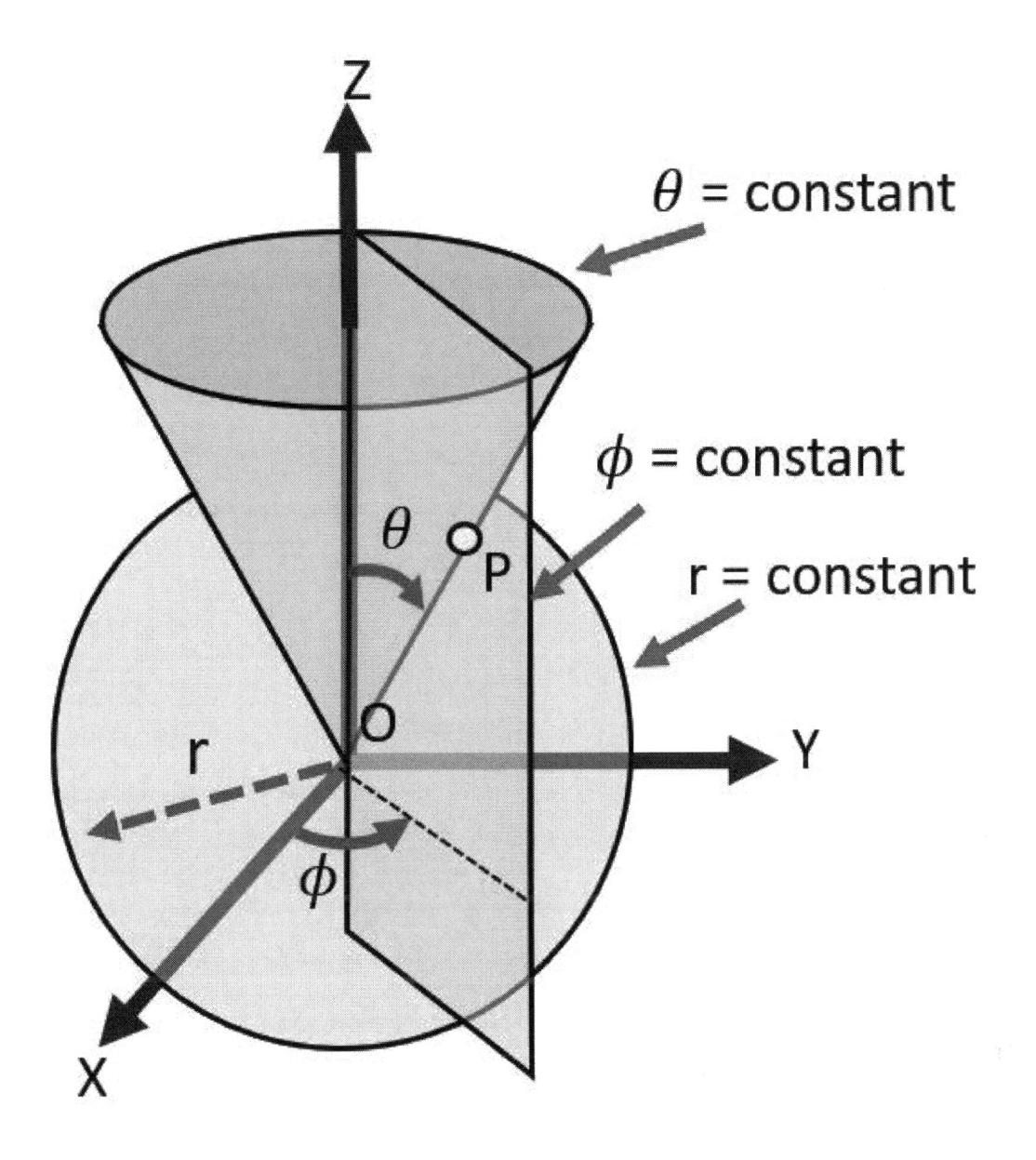

A point in space in Spherical coordinate system is specified in terms of,

- the distance of the point from the origin (r)
- the angle the line joining the point to the origin makes with the z- axis (θ).
- the angle a half plane containing the point makes with the x- axis in the anticlockwise direction (ϕ) (same as the azimuth angle in cylindrical coordinates)

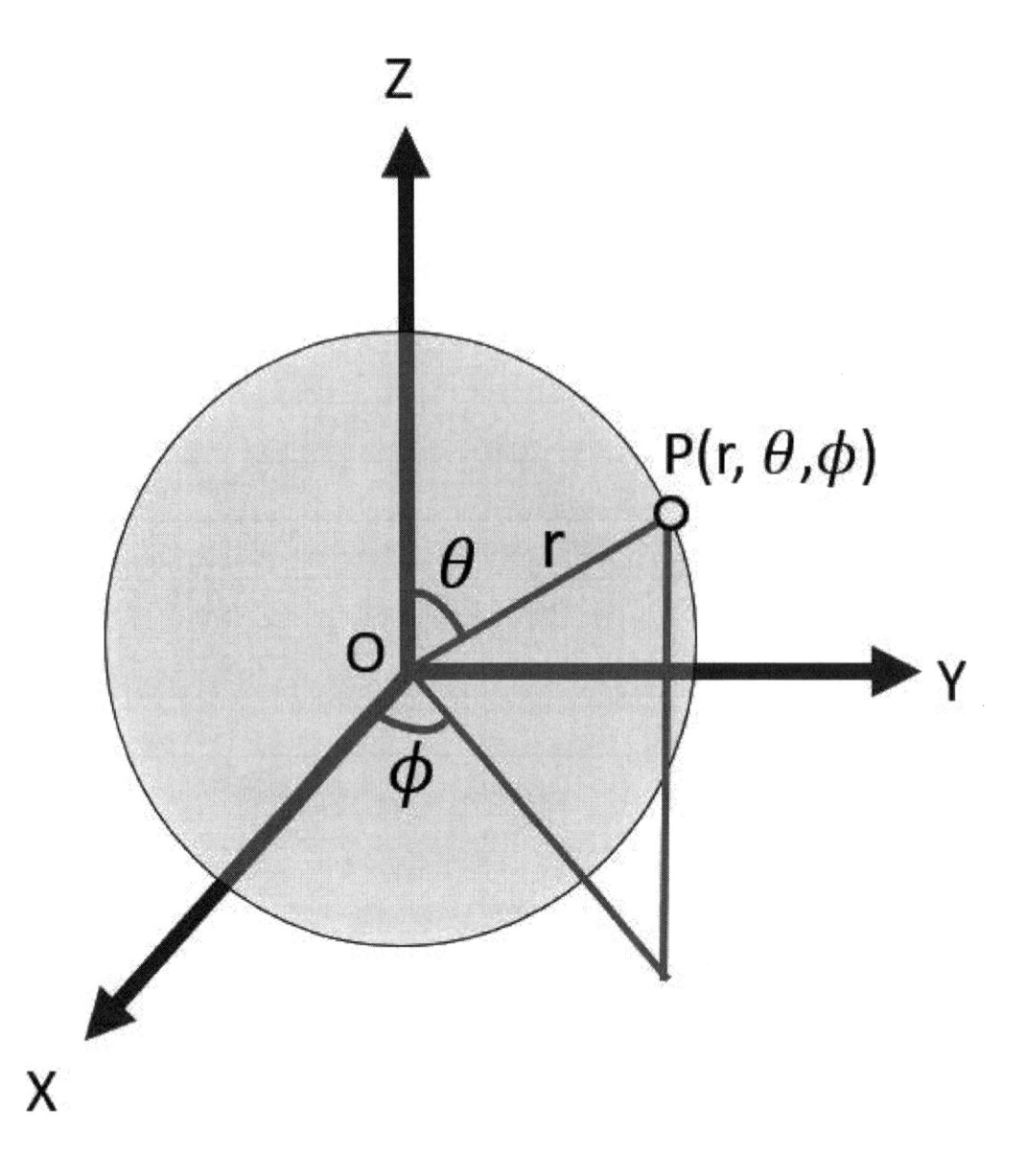

To cover the entire space, the radius of the sphere (r) can be varied from 0 to ∞, the slant angle of the cone (θ) can be varied from 0 to π radians, and the azimuth angle (ϕ) can be varied from 0 to 2π radians.

Spherical to Cartesian coordinates:

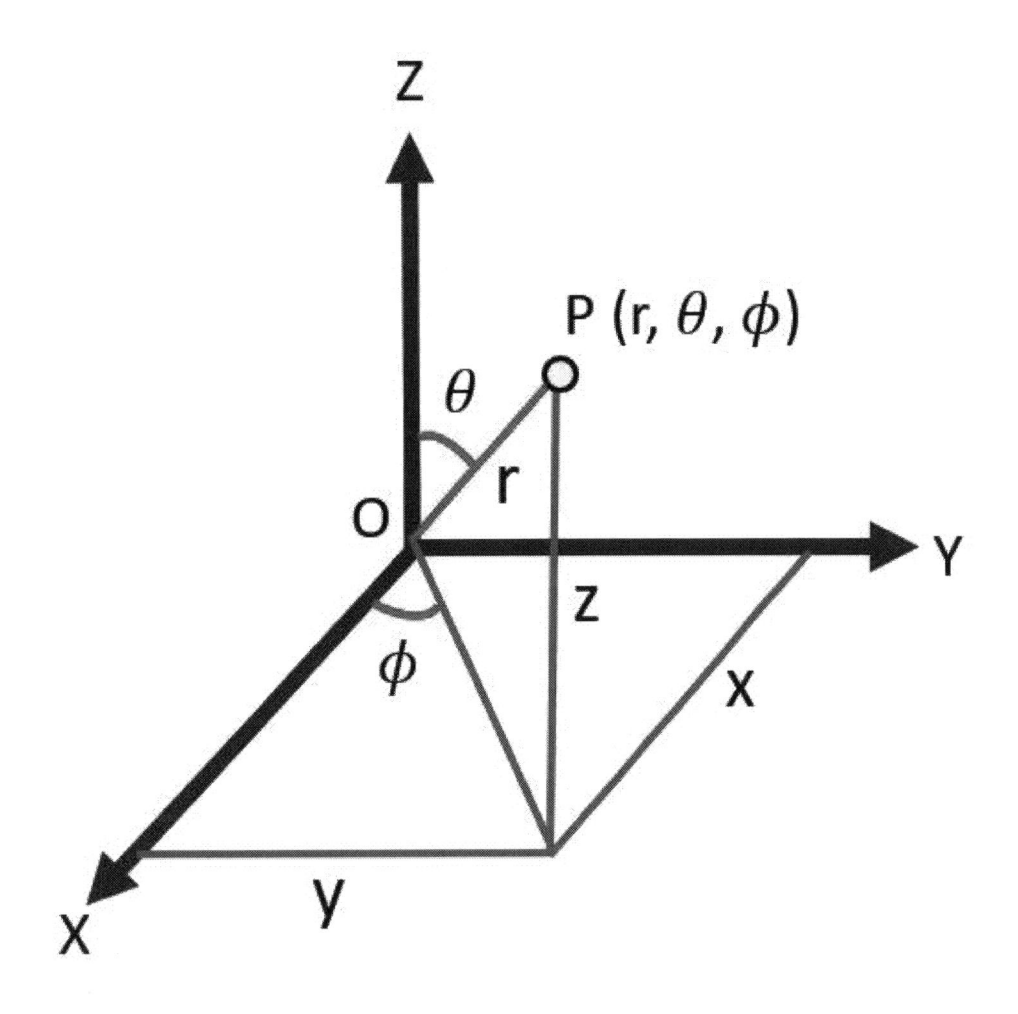

Spherical to Cartesian

$x = r \sin\theta \cos\phi$
$y = r \sin\theta \sin\phi$
$z = r \cos\theta$

$\Leftrightarrow$

$r = \sqrt{x^2 + y^2 + z^2}$
$\theta = \tan^{-1}(\sqrt{x^2 + y^2}/z)$
$\phi = \tan^{-1}(y/x)$

Unit Vector in Spherical coordinates:

Unit vectors

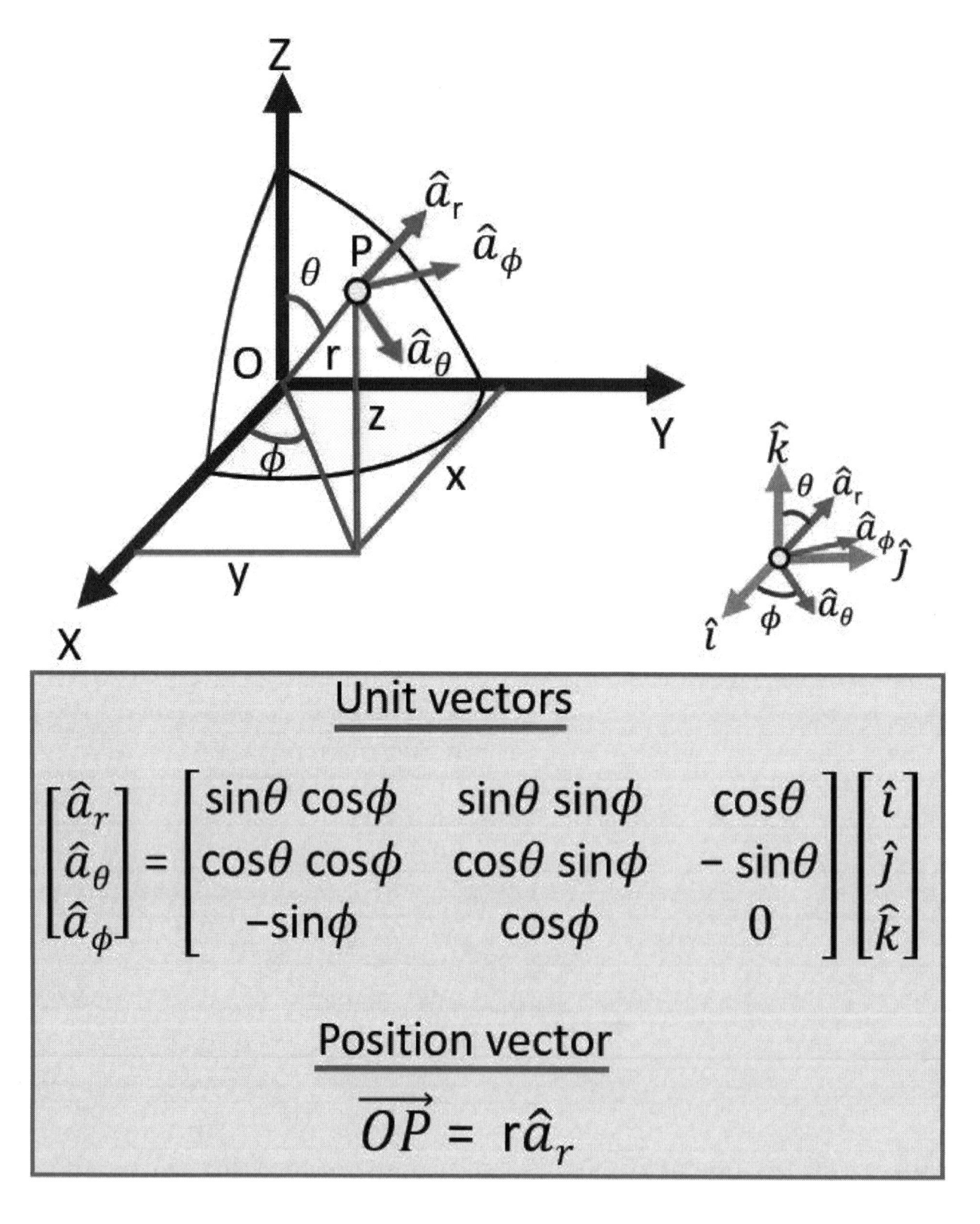

$$\begin{bmatrix} \hat{a}_r \\ \hat{a}_\theta \\ \hat{a}_\phi \end{bmatrix} = \begin{bmatrix} \sin\theta\cos\phi & \sin\theta\sin\phi & \cos\theta \\ \cos\theta\cos\phi & \cos\theta\sin\phi & -\sin\theta \\ -\sin\phi & \cos\phi & 0 \end{bmatrix} \begin{bmatrix} \hat{i} \\ \hat{j} \\ \hat{k} \end{bmatrix}$$

Position vector

$$\overrightarrow{OP} = r\hat{a}_r$$

Hence, in spherical coordinates any vector A can be expressed in terms of the fundamental unit vectors as,

$$\vec{A} = A_r \hat{a}_r + A_\theta \hat{a}_\theta + A_\phi \hat{a}_\phi$$

Just like in cylindrical coordinates, in spherical coordinates too the fundamental unit vectors are not universal. Meaning the fundamental unit vectors at different points are different. Therefore, the derivatives of these unit vectors are non zero.

$\dfrac{\partial(\hat{a}_r)}{\partial r} = 0$	$\dfrac{\partial(\hat{a}_\theta)}{\partial r} = 0$	$\dfrac{\partial(\hat{a}_\phi)}{\partial r} = 0$
$\dfrac{\partial(\hat{a}_r)}{\partial \theta} = \hat{a}_\theta$	$\dfrac{\partial(\hat{a}_\theta)}{\partial \theta} = -\hat{a}_r$	$\dfrac{\partial(\hat{a}_\phi)}{\partial \theta} = 0$
$\dfrac{\partial(\hat{a}_r)}{\partial \phi} = \sin\theta\, \hat{a}_\phi$	$\dfrac{\partial(\hat{a}_\theta)}{\partial \phi} = \cos\theta\, \hat{a}_\phi$	$\dfrac{\partial(\hat{a}_\phi)}{\partial \phi} = -\sin\theta\, \hat{a}_r - \cos\theta\, \hat{a}_\theta$

Differential Length, Volume in Spherical coordinates:

To find differential length in spherical coordinates, we need to consider the differential change in any one variable, keeping the other variables unchanged.

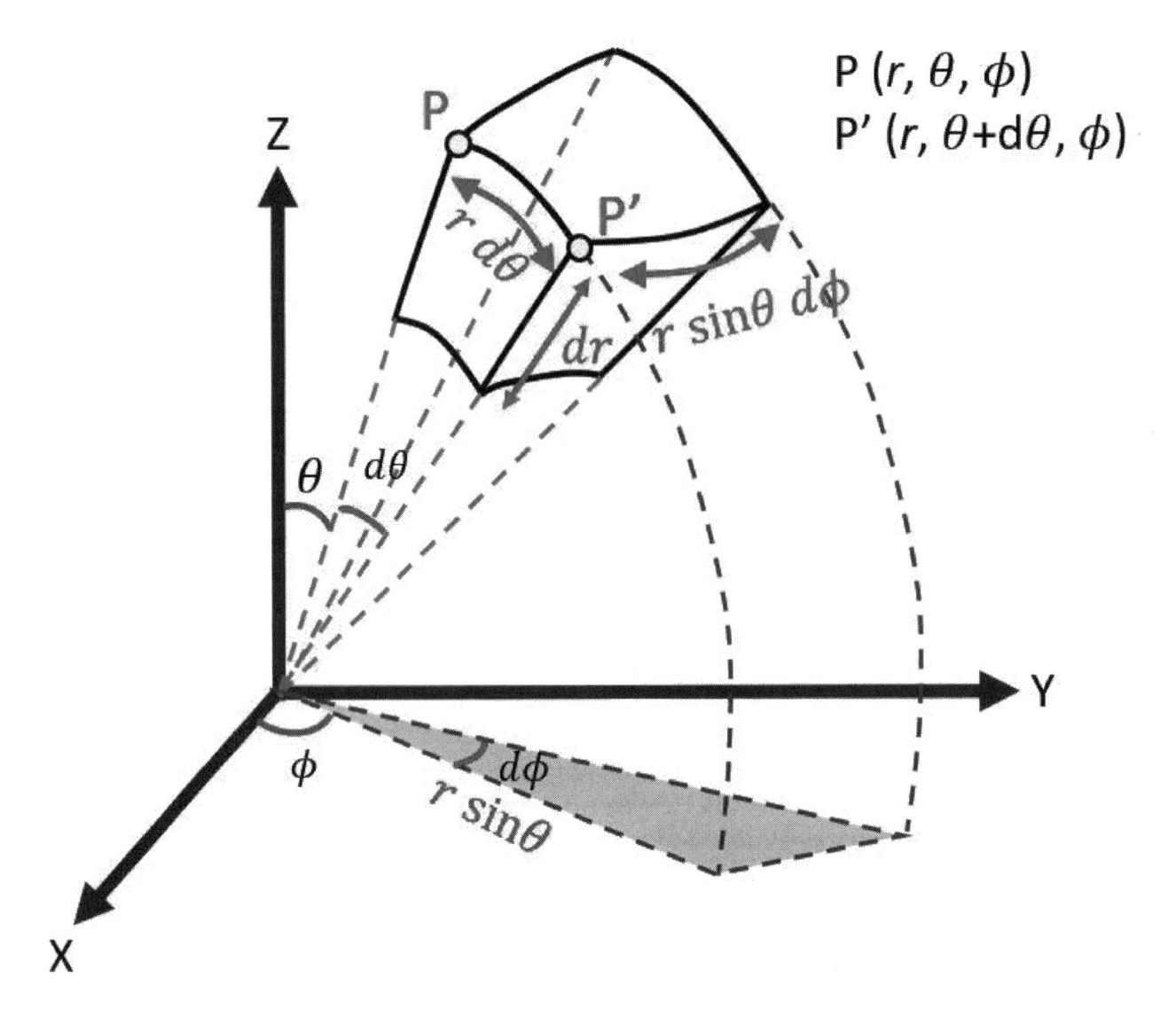

For instance, to calculate the partial derivative with respect to θ, assume a point moves from P(ρ, θ, z) to P'(ρ, θ+dθ, z) keeping variables ρ and **z** constant. Hence the differential length is equal to the length of the arc PP' i.e. **dl** = $\rho d\theta$. Similarly the differential change along **r** and ϕ directions are dr and $d\phi$ respectively.

Therefore, in general the differential length vector **dl** is given by,

$$\vec{dl} = dr\ \hat{a}_r + rd\theta\ \hat{a}_\theta + r\sin\theta d\phi\ \hat{a}_\phi$$

The volume of the differential element is given by,

142

$$dV = dr\,(r d\theta)\,(r\sin\theta\, d\phi)$$
$$= r^2 \sin\theta\, dr\, d\theta\, d\phi$$

Del Operator in Spherical coordinates:

The Del operator in the Spherical coordinates is given by,

$$\nabla = \frac{\partial}{\partial r}\hat{a}_r + \frac{1}{r}\frac{\partial}{\partial \theta}\hat{a}_\theta + \frac{1}{r\sin\theta}\frac{\partial}{\partial \phi}\hat{a}_\phi$$

Gradient:

The expression for gradient in spherical coordinates is,

$$\nabla f = \frac{\partial f}{\partial r}\hat{a}_r + \frac{1}{r}\frac{\partial f}{\partial \theta}\hat{a}_\theta + \frac{1}{r\sin\theta}\frac{\partial f}{\partial \phi}\hat{a}_\phi$$

It goes without mentioning that the scalar field **f** should also be in spherical coordinates.

Divergence:

The divergence of a vector field in spherical coordinates is given by,

$$\nabla \cdot \vec{A} = \frac{1}{r^2}\frac{\partial (r^2 A_r)}{\partial r} + \frac{1}{r\sin\theta}\frac{\partial (\sin\theta\, A_\theta)}{\partial \theta} + \frac{1}{r\sin\theta}\frac{\partial (A_\phi)}{\partial \phi}$$

Curl:

The curl of a vector field in spherical coordinates can be obtained as,

$$\nabla \times \vec{A} = \frac{1}{r^2 \sin\theta} \begin{vmatrix} \hat{a}_r & r\hat{a}_\theta & r\sin\theta\, \hat{a}_\phi \\ \frac{\partial}{\partial r} & \frac{\partial}{\partial \theta} & \frac{\partial}{\partial \phi} \\ A_r & rA_\theta & r\sin\theta\, A_\phi \end{vmatrix}$$

Laplacian:

The Laplacian in spherical coordinates is given by,

$$\nabla^2 f = \frac{1}{r^2}\frac{\partial}{\partial r}\left(r^2 \frac{\partial f}{\partial r}\right) + \frac{1}{r^2 \sin\theta}\frac{\partial}{\partial \theta}\left(\sin\theta \frac{\partial f}{\partial \theta}\right) + \frac{1}{r^2 \sin^2\theta}\frac{\partial^2 f}{\partial \phi^2}$$

The derivations of the above expressions are too cumbersome, so we have not included them here. But try them out if you are interested. The trick is to expand all the terms and use product rule wherever the fundamental unit vectors are involved.

Solved Examples:

8. **Convert the point P(1,1,3) from cartesian to spherical coordinates.**

$$r = \sqrt{x^2 + y^2 + z^2} = \sqrt{1^2 + 1^2 + 3^2} = 3.31$$
$$\theta = \tan^{-1}(\sqrt{x^2 + y^2}/z) = \tan^{-1}(\sqrt{2}/3) = 25.2°$$
$$\phi = \tan^{-1}(y/x) = \tan^{-1}(1/1) = 45°$$

Therefore in spherical coordinates the point is P(3.31, 25.2°, 45°).

9. **Convert the point P(7.07, 45°, 53.1°) from spherical to cartesian coordinates.**

$$x = r\sin\theta\cos\phi = 7.07\sin 45\cos 53.1 = 3$$

$$y = r\sin\theta\sin\phi = 7.07\sin 45\sin 53.1 = 4$$

$$z = r\cos\theta = 7.07\cos 45 = 5$$

Therefore in cartesian coordinates the point is P(3,4,5).

10.   A vector field is given as A = a$_r$/r in spherical coordinates. Convert the vector field to cartesian coordinates.

$$\vec{A} = \frac{\hat{a}_r}{r}$$

$r = \sqrt{x^2 + y^2 + z^2}$

Expressing x, y, z in spherical coordinates

$$x = r\sin\theta\cos\phi$$
$$y = r\sin\theta\sin\phi$$
$$z = r\cos\theta$$

Expressing $\hat{a}_r$ in spherical coordinates

$$\hat{a}_r = \sin\theta\cos\phi\,\hat{i} + \sin\theta\sin\phi\,\hat{j} + \cos\theta\,\hat{k}$$

$$\therefore \vec{A} = \frac{\sin\theta\cos\phi\,\hat{i} + \sin\theta\sin\phi\,\hat{j} + \cos\theta\,\hat{k}}{r}$$

$$= \frac{x\hat{i} + y\hat{j} + z\hat{k}}{r^2}$$

$$= \frac{x\hat{i} + y\hat{j} + z\hat{k}}{x^2+y^2+z^2}$$

11.

Find the gradient of the scalar function $f = -A(r + a^3/2r^2)\cos\theta$.

$$f = -A\left(r + \frac{a^3}{2r^2}\right)\cos\theta$$

$$\nabla f = \frac{\partial f}{\partial r}\hat{a}_r + \frac{1}{r}\frac{\partial f}{\partial \theta}\hat{a}_\theta + \frac{1}{r\sin\theta}\frac{\partial f}{\partial \phi}\hat{a}_\phi$$

$$\frac{\partial f}{\partial r} = -A\cos\theta\left(1 - \frac{a^3}{r^3}\right)$$

$$\frac{\partial f}{\partial \theta} = A\sin\theta\left(r + \frac{a^3}{2r^2}\right)$$

$$\frac{\partial f}{\partial \phi} = 0$$

$$\nabla f = -A\cos\theta\left(1 - \frac{a^3}{r^3}\right)\hat{a}_r + A\sin\theta\left(1 + \frac{a^3}{2r^3}\right)\hat{a}_\theta$$

12.

Obtain the divergence of the vector field $F = \langle \cos\theta/r^2,\ r\sin\theta\cos\phi,\ \cos\theta \rangle$.

$$\vec{F} = \frac{1}{r^2}\cos\theta \,\hat{a}_r + r\sin\theta \cos\phi \,\hat{a}_\theta + \cos\theta \,\hat{a}_\phi$$

$$\nabla \cdot \vec{F} = \frac{1}{r^2}\frac{\partial(r^2 F_r)}{\partial r} + \frac{1}{r\sin\theta}\frac{\partial(\sin\theta \,F_\theta)}{\partial \theta} + \frac{1}{r\sin\theta}\frac{\partial(F_\phi)}{\partial \phi}$$

$$\frac{\partial(r^2 F_r)}{\partial r} = \frac{\partial(r^2 \times \frac{1}{r^2}\cos\theta)}{\partial r} = 0$$

$$\frac{\partial(\sin\theta \,F_\theta)}{\partial \theta} = \frac{\partial(r\sin^2\theta \cos\phi)}{\partial \theta} = 2r\sin\theta \cos\theta \cos\phi$$

$$\frac{\partial(F_\phi)}{\partial \phi} = \frac{\partial(\cos\theta)}{\partial \phi} = 0$$

$$\therefore \nabla \cdot \vec{F} = \frac{2r\sin\theta \cos\theta \cos\phi}{r\sin\theta} = 2\cos\theta \cos\phi$$

13.

Obtain the curl of the vector field $\vec{F} = \langle \cos\theta/r^2,\ r\sin\theta\cos\phi,\ \cos\theta \rangle$.

$$\vec{F} = \frac{1}{r^2}\cos\theta\ \hat{a}_r + r\sin\theta\cos\phi\ \hat{a}_\theta + \cos\theta\ \hat{a}_\phi$$

$$\nabla \times \vec{F} = \frac{1}{r^2 \sin\theta} \begin{vmatrix} \hat{a}_r & r\hat{a}_\theta & r\sin\theta\ \hat{a}_\phi \\ \frac{\partial}{\partial r} & \frac{\partial}{\partial \theta} & \frac{\partial}{\partial \phi} \\ F_r & rF_\theta & r\sin\theta F_\phi \end{vmatrix}$$

$$= \frac{1}{r^2 \sin\theta}\left[\left(\frac{\partial(r\sin\theta F_\phi)}{\partial \theta} - \frac{\partial(rF_\theta)}{\partial \phi}\right)\hat{a}_r + \right.$$
$$\left(\frac{\partial(F_r)}{\partial \phi} - \frac{\partial(r\sin\theta F_\phi)}{\partial r}\right) r\hat{a}_\theta +$$
$$\left.\left(\frac{\partial(rF_\theta)}{\partial r} - \frac{\partial(F_r)}{\partial \theta}\right) r\sin\theta\ \hat{a}_\phi\right]$$

$$= \frac{1}{r^2 \sin\theta}\left[\left(\frac{\partial(r\sin\theta\cos\theta)}{\partial \theta} - \frac{\partial(r^2\sin\theta\cos\phi)}{\partial \phi}\right)\hat{a}_r + \right.$$
$$\left(\frac{\partial(\frac{1}{r^2}\cos\theta)}{\partial \phi} - \frac{\partial(r\sin\theta\cos\theta)}{\partial r}\right) r\hat{a}_\theta +$$
$$\left.\left(\frac{\partial(r^2\sin\theta\cos\phi)}{\partial r} - \frac{\partial(\frac{1}{r^2}\cos\theta)}{\partial \theta}\right) r\sin\theta\ \hat{a}_\phi\right]$$

$$= \frac{1}{r^2 \sin\theta}\left[(r\cos 2\theta + r^2\sin\theta\sin\phi)\hat{a}_r - (\sin\theta\cos\theta)\ r\hat{a}_\theta + \right.$$
$$\left.\left(2r\sin\theta\cos\phi + \frac{1}{r^2}\sin\theta\right) r\sin\theta\ \hat{a}_\phi\right]$$

$$= \left(\frac{\cos 2\theta}{r\sin\theta} + \sin\phi\right)\hat{a}_r - \left(\frac{\cos\theta}{r}\right)\hat{a}_\theta + \left(2\cos\phi + \frac{1}{r^3}\sin\theta\right)\sin\theta\ \hat{a}_\phi$$

14.

Find the gradient of the scalar function f = A rcosθ.

$$f = A\, r\cos\theta$$

$$\nabla^2 f = \frac{1}{r^2}\frac{\partial}{\partial r}\left(r^2 \frac{\partial f}{\partial r}\right) + \frac{1}{r^2 \sin\theta}\frac{\partial}{\partial \theta}\left(\sin\theta \frac{\partial f}{\partial \theta}\right) + \frac{1}{r^2 \sin^2\theta}\frac{\partial^2 f}{\partial \phi^2}$$

$$r^2 \frac{\partial f}{\partial r} = r^2 A \cos\theta$$

$$\frac{1}{r^2}\frac{\partial}{\partial r}\left(r^2 \frac{\partial f}{\partial r}\right) = \frac{2A \cos\theta}{r}$$

$$\sin\theta \frac{\partial f}{\partial \theta} = -A\, r\sin^2\theta$$

$$\frac{1}{r^2 \sin\theta}\frac{\partial}{\partial \theta}\left(\sin\theta \frac{\partial f}{\partial \theta}\right) = \frac{-2A \cos\theta}{r}$$

$$\frac{1}{r^2 \sin^2\theta}\frac{\partial^2 f}{\partial \phi^2} = 0$$

$$\therefore \nabla^2 f = 0$$

7. APPLICATIONS

Vectors have lots of practical applications in physics and engineering. In this chapter we look at some of those. We won't go into depth on any of the topics, our main focus will be on problem solving and thereby, to give you an idea about the kind of problems where vectors are used in real life.

7.1 MECHANICS

Mechanics is a branch of physics that deals bodies at rest or in motion under the influence of forces. The word "Mechanics" is derived from the greek word "mechane", which means machine. It is one of the oldest, if not the oldest branches of physical science. The main objective of mechanics is to analyze and predict the physical behavior of objects and thereby make them suitable for engineering applications.

Displacement:

Displacement is defined as the change in position of an object. It is a vector quantity whose magnitude is the shortest distance from the initial to the final position of an object undergoing motion. It's denoted as **s** (or **x** in some cases) and its unit is metres.

1. **A bear travels 70 kms in a north-east direction from his den. It then travels 150 kms 60 degrees north of west. Determine how far and in what direction the bear is from his den.**

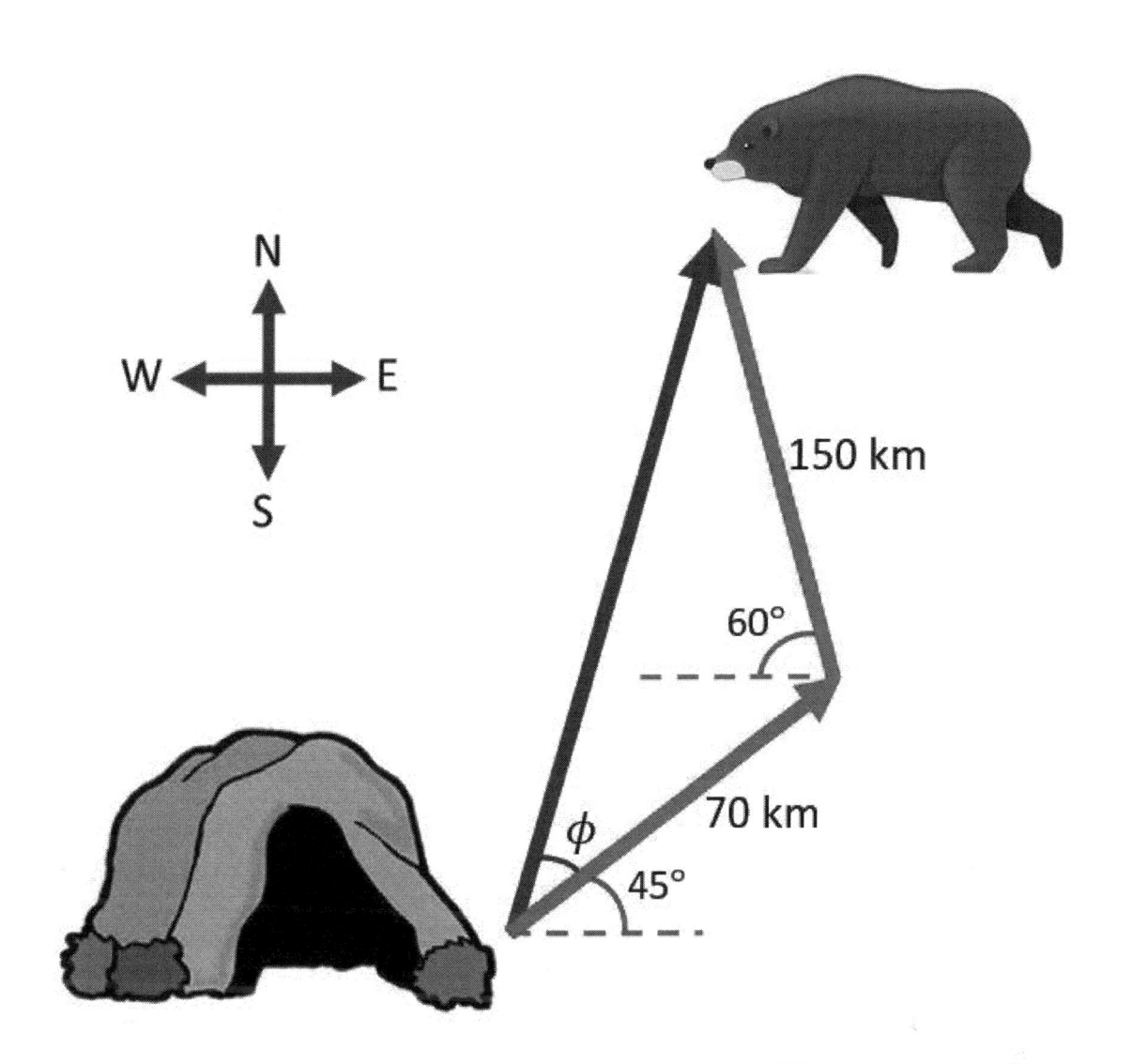

$$s = \sqrt{s_1^2 + s_2^2 + 2s_1 s_2 \cos\theta}$$
$$= \sqrt{70^2 + 150^2 + 2(70)(150)\cos 75}$$
$$= 181.2 \text{ km}$$

$$\phi = \tan^{-1}\left(\frac{s_2 \sin\theta}{s_1 + s_2 \cos\theta}\right)$$
$$= \tan^{-1}\left(\frac{150 \sin 75}{70 + 150 \cos 75}\right)$$
$$= 53.09°$$

Velocity:

The velocity of an object is defined as the rate of change of its position (displacement) with respect to time i.e. $\mathbf{v} = \frac{ds}{dt}$. It's a vector quantity and its unit is metres/second (m/s).

2. A Boat has a velocity of 5m/s and it aims to cross over to point A. However, because of the river current it reaches another point B. If the velocity of the current is 2m/s, then calculate

   a. the resultant velocity of the motorboat
   b. the time taken to travel shore to shore.
   c. the distance between points A and B.

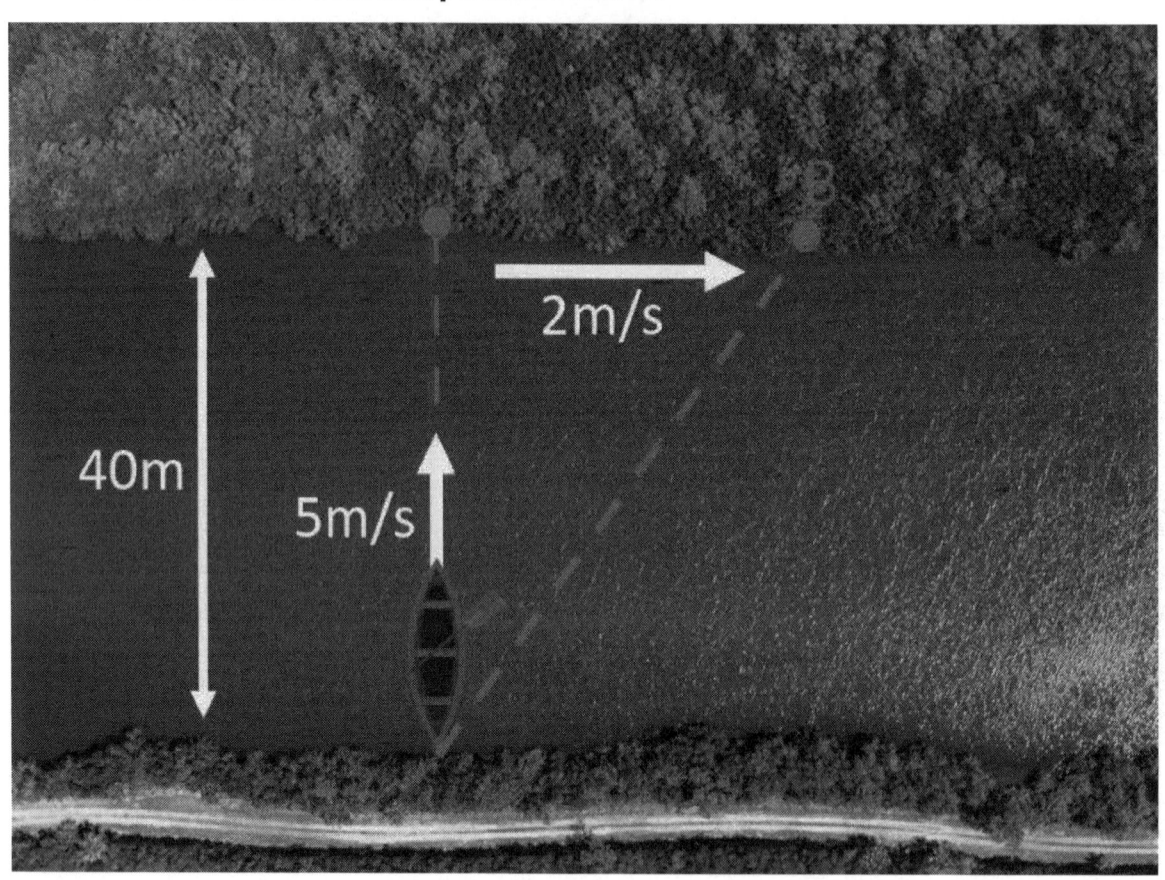

The resultant velocity can be found out using simple vector addition as,

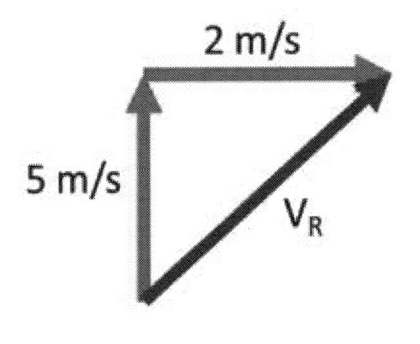

$$V_R = \sqrt{5^2 + 2^2}$$
$$= 5.38 \text{ m/s}$$

To answer the next parts (b & c) of the problem, consider the motion in 2 parts; from shore to shore and along the river.

b) Shore to shore motion:
s = width of the river = 40m
v = velocity of the boat = 5m/s

Time taken $t = \dfrac{s}{v} = 8$ seconds

c) Motion along the river:
s = Distance between point A & B
v = velocity of the current = 2m/s

Distance between A and B
$s = vt = 16$ m

While the problem itself is very easy, the difficult bit is using the right quantities in the appropriate part of the problem. For example, a lot of times students make the mistake of using the resultant velocity in the expression to calculate the time taken (in the above problem). But that obviously is incorrect. While considering the shore to shore motion only use those quantities, similarly for the other. The only quantity that remains the same is the time taken. Suppose the diagonal distance had been given, in that case the time taken can be calculated as t = diagonal distance/resultant velocity.

Acceleration:

Acceleration is defined as the rate of change of velocity of an object with respect to time i.e. $a = \frac{dv}{dt}$. It's a vector quantity and its unit is m/s². Displacement, velocity, acceleration and time are interconnected through 3 equations called the equations of motion.

- $v = u + at$
- $s = ut + \frac{1}{2}at^2$
- $v^2 - u^2 = 2as$

u = initial velocity
v = final velocity
s = displacement
a = acceleration
t = time

3. A block slides down an incline of slope $\theta = 36.86°$ due to its own weight. What is the acceleration of the block? (Take g = 9.8 m/s²)

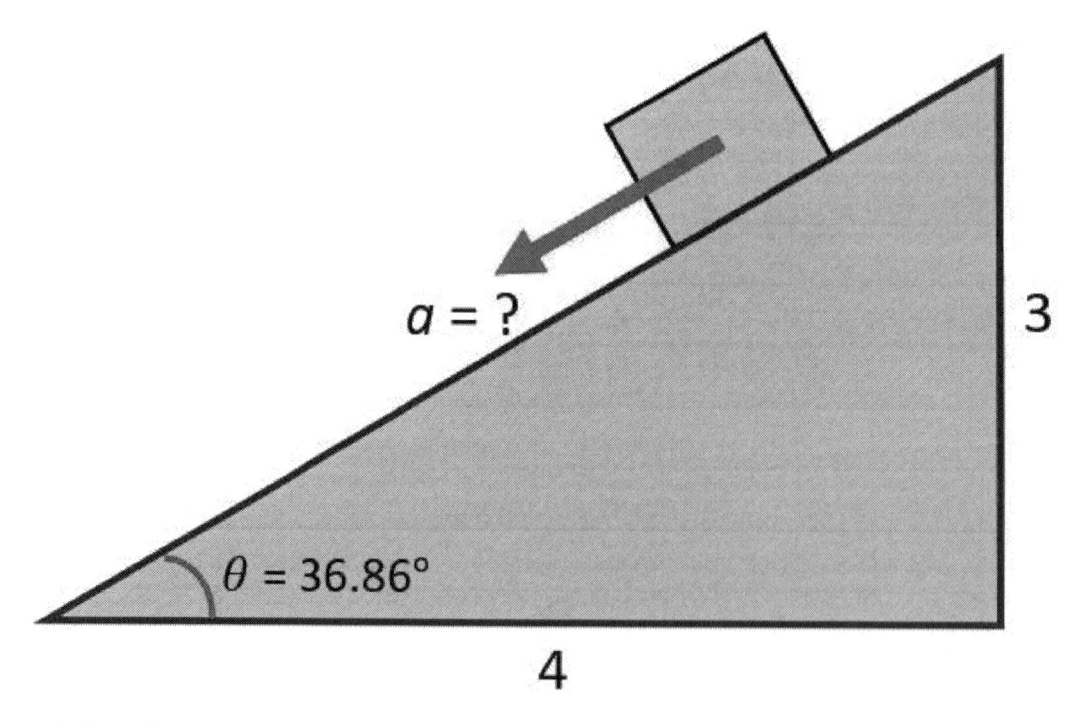

Resolving the acceleration due to gravity **g** into 2 components (along the incline and normal to it), we can obtain the acceleration **a** as,

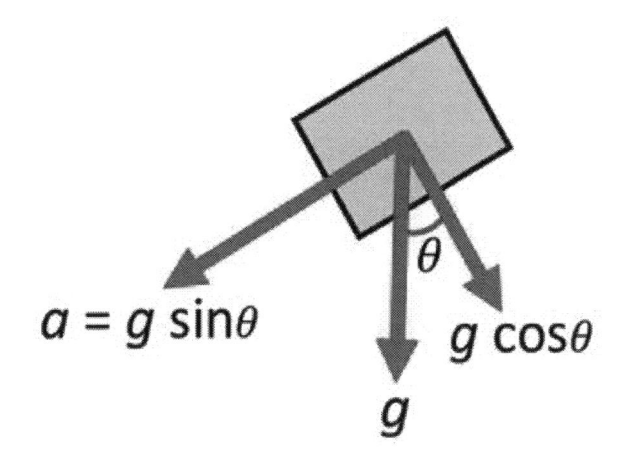

$$a = g \sin \theta$$
$$= g \sin(36.86°)$$
$$= \frac{3g}{5} = 5.88 \text{ m/s}^2$$

4. A football is kicked with an initial velocity of 25m/s at an angle 45 degrees with the ground. Determine

   a. The peak height the football attains.
   b. Time of flight.
   c. Horizontal displacement (Range).

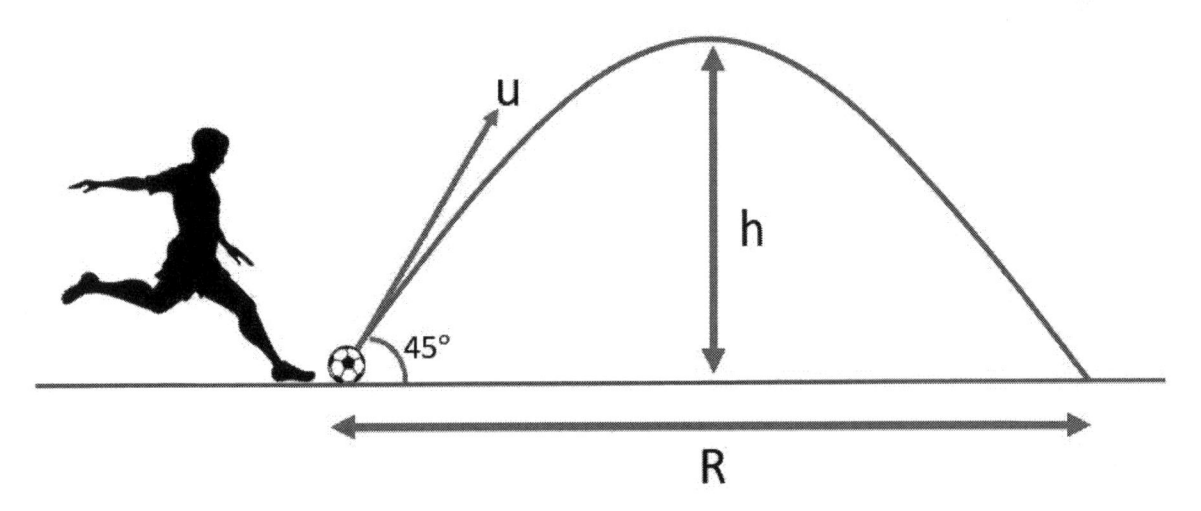

In this problem too we have to consider the motion in 2 parts; along the vertical and the horizontal direction.

Vertical motion:
h = max height
u = initial velocity upwards
  = 25 sin45 = 17.68 m/s
a = -g = -9.8 m/s^2

Horizontal motion:
R = range
v = initial velocity horizontally
  = 25 cos45 = 17.68 m/s

When the football reaches the max height, it has zero velocity in the vertical direction.

$$v^2 - u^2 = 2as$$
$$0 - 17.68^2 = 2(-9.8)h$$
$$\Rightarrow h = 15.95 \text{ m}$$

Time of flight is twice the time taken by the football to reach the maximum height.

$$v - u = at$$
$$0 - 17.68 = -9.8 t_{up}$$
$$\Rightarrow t_{up} = 1.8 \text{ s}$$
$$\therefore t = 2 \times t_{up} = 3.6 \text{ s}$$

Range of the projectile motion can be obtained as,

$$s = ut + at^2/2$$
$$R = 17.68 \times 3.6$$
$$\Rightarrow R = 63.648 \text{ m}$$

Force:

In mechanics, Force is defined as any action (push or a pull) that tends to maintain or alter the motion of a body or to distort it. It is the product of the mass and the acceleration of an object (F= ma). It is a vector quantity and its unit is Newton.

For objects at rest, all forces acting on it are balanced in all directions.

5. **Find the tension T_1 & T_2 in the strings in the figure shown below. (Use g = 10)**

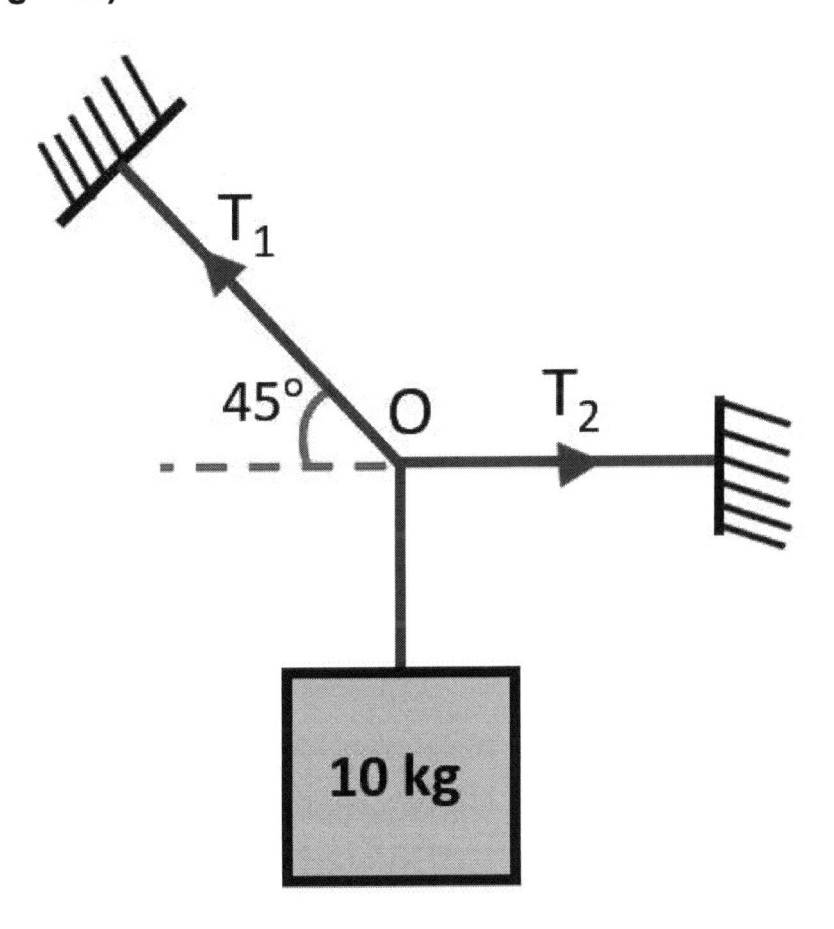

The tension T_1 can be resolved into the horizontal and vertical components as shown.

$T_1 \sin 45$

T_1

$45°$

$T_1 \cos 45$ ← →T_2

↓

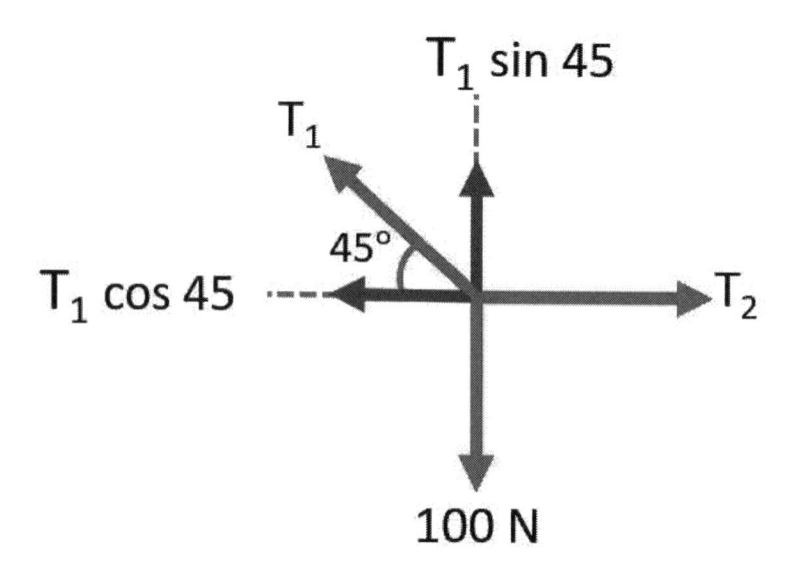

100 N

The force acting due to the 10kg weight is 10g = 100N. Since the block is in equilibrium, the horizontal and the vertical forces must be balanced.

For vertical forces,

$$T_1 \sin 45 = 100$$
$$\Rightarrow T_1/\sqrt{2} = 100$$
$$\Rightarrow T_1 = 100\sqrt{2} = \underline{\underline{141.42 \text{ N}}}$$

For horizontal forces,

$$T_1 \cos 45 = T_2$$
$$\Rightarrow T_2 = (100\sqrt{2})/\sqrt{2}$$
$$= \underline{\underline{100 \text{ N}}}$$

6. Two masses of 4 kg and 5 kg are connected by a string passing through a frictionless pulley and the 4kg mass is kept on a smooth table as shown in the figure. The acceleration of 5 kg mass is?

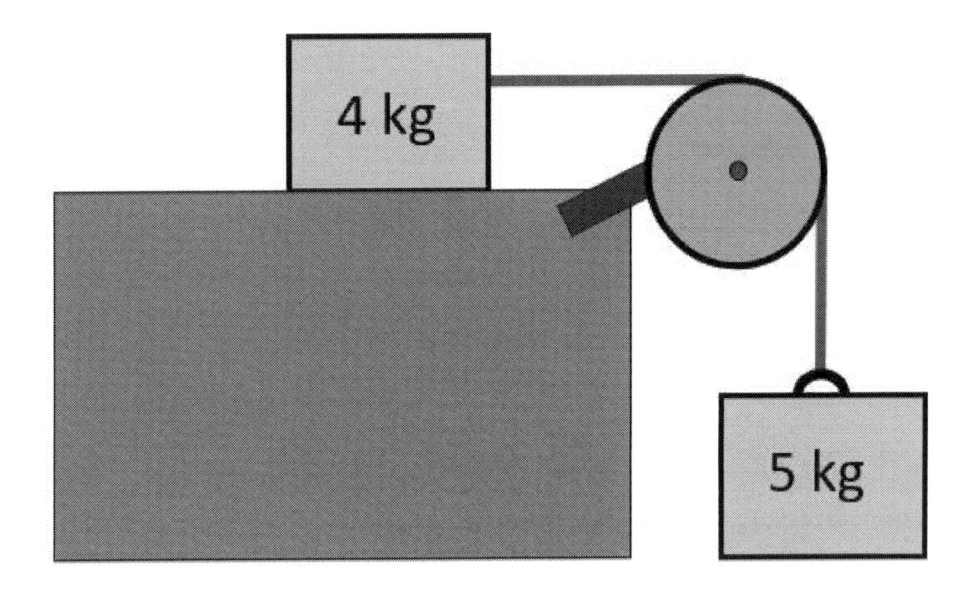

The first step in solving such problems is to draw the free body diagrams and write the corresponding force equations.

5kg Block:

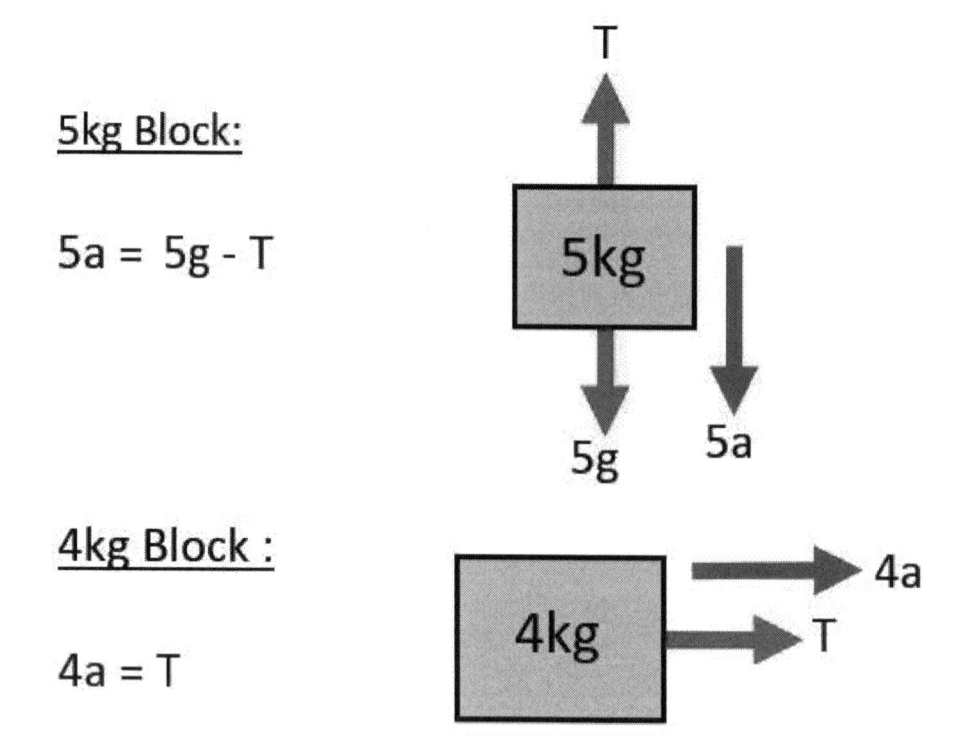

$5a = 5g - T$

4kg Block:

$4a = T$

Solving these equations,

$$5a = 5g - 4a$$
$$\Rightarrow a = \frac{5g}{9} = 5.44 m/s^2$$

7. Two blocks with mass m_1 and mass m_2 are hung on a pulley system as shown in the figure. Find the magnitude of the acceleration with which the blocks m_1 and m_2 are moving and the magnitude of the tension force T in the rope. Ignore the masses of the pulley system and the rope.

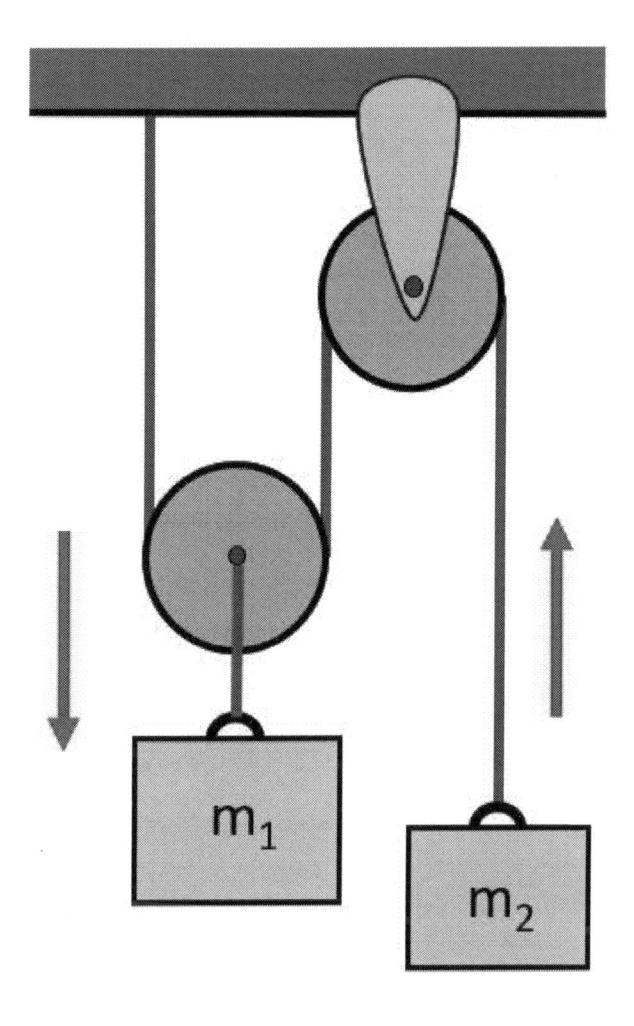

In this problem also we start by drawing the free body diagrams and obtaining the force equations.

Block m_2:

$m_2 a_2 = T - m_2 g$

Block m_1:

$m_1 a_1 = m_1 g - 2T$

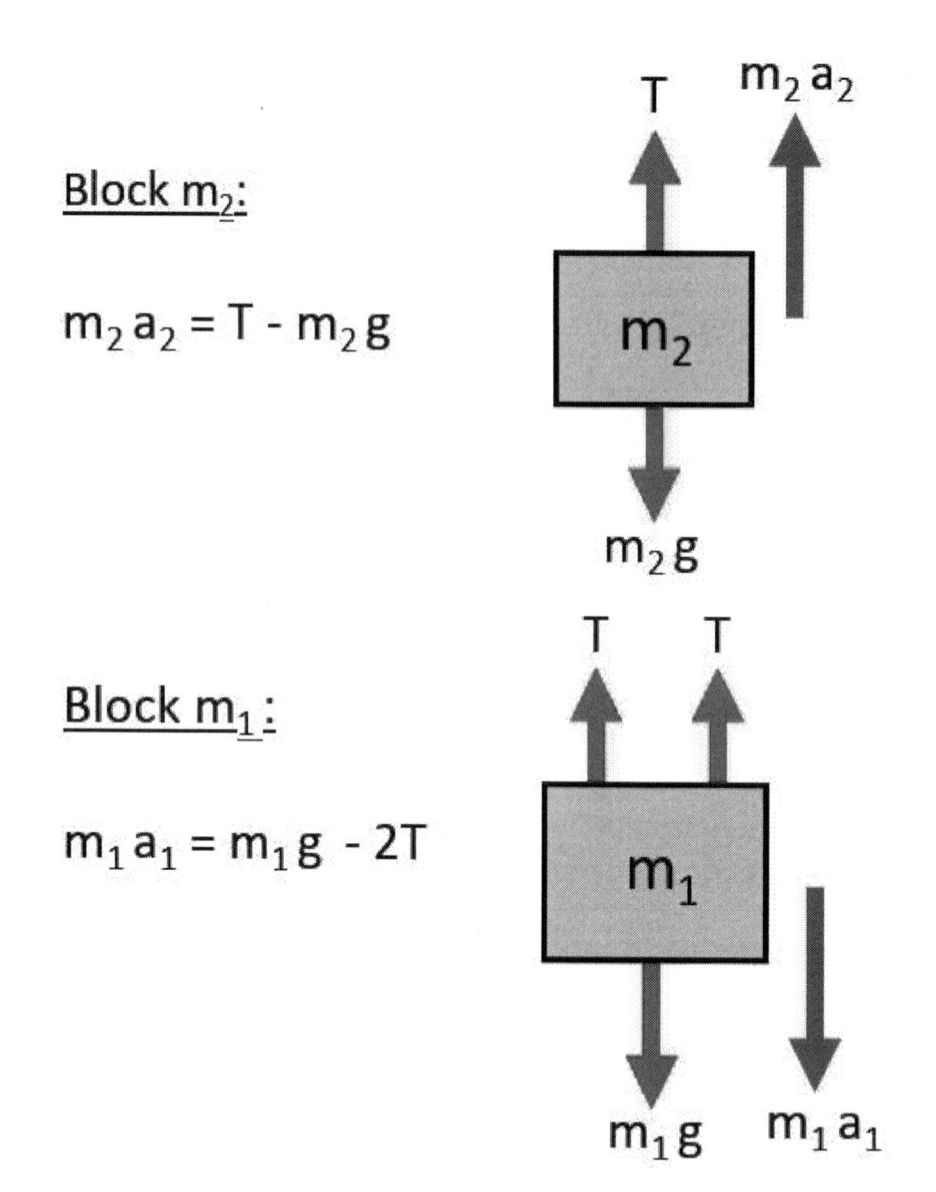

In such a pulley arrangement there is an interesting relationship between the accelerations **a₁** and **a₂**.

$$a_2 = 2a_1$$

$$\therefore m_2 \, 2a_1 = T - m_2 g \; \&$$
$$m_1 a_1 = m_1 g - 2T$$

Solving these equations,

$$a_1 = \frac{(m_1 - 2m_2)g}{m_1 + 4m_2} \; \&$$

$$a_2 = 2\frac{(m_1 - 2m_2)g}{m_1 + 4m_2}$$

$$T = m_2 \, 2a_1 - m_2 g$$
$$= \frac{-3m_1 m_2 g}{m_1 + 4m_2}$$

8. **Determine the force in each member of the truss shown in the figure and indicate whether the members are in tension or compression**

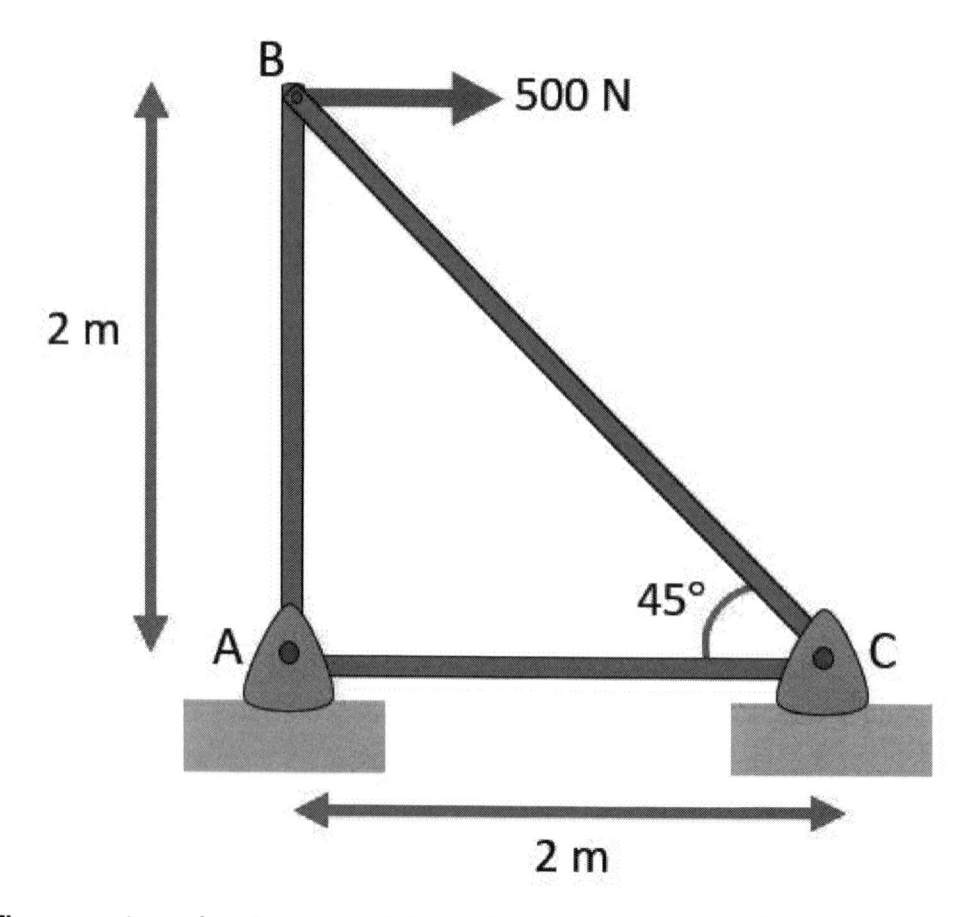

The way to solve truss problems is to consider each joint separately and solve equations of equilibrium for each.

Joint B:

$\sum F_x = 0$: $500 - F_{BC} \sin 45° = 0$
$\Rightarrow F_{BC} = 707.1$ N (C)

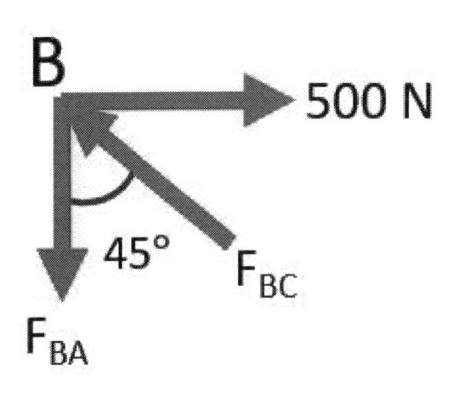

$\sum F_y = 0$: $F_{BC} \cos 45° - F_{BA} = 0$
$\Rightarrow F_{BA} = 500$ N (T)

Joint C:

$\sum F_x = 0$: $-F_{CA} + F_{BC} \cos 45° = 0$
$\Rightarrow F_{CA} = 500$ N (T)

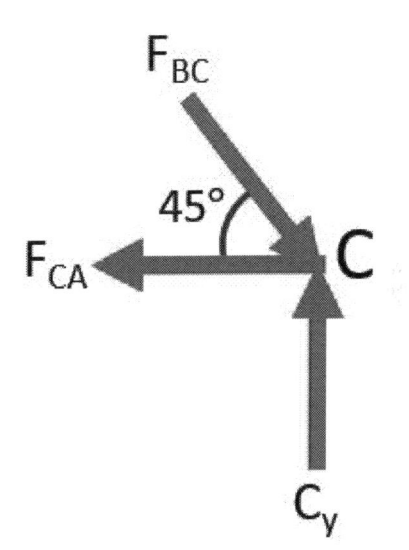

$\sum F_y = 0$: $C_y - F_{BC} \sin 45° = 0$
$\Rightarrow C_y = 500$ N

7.2 ELECTROMAGNETISM

Electromagnetism is a branch of physics involving the study of the electromagnetic force, a type of physical interaction that occurs between electrically charged particles.

Electric Field Intensity:

The electric field Intensity at a point is the force experienced by a unit positive charge placed at that point due to the presence of another charge in its vicinity. It is a vector quantity and its unit is N/C.

$$\vec{E} = k q \frac{(\vec{r} - \vec{r}_1)}{|\vec{r} - \vec{r}_1|^3}$$

9. **Three equal positive charges of 4 Coulomb each are placed at the 3 corners of a square of side 0.2m. Determine the magnitude & direction of the electric field intensity at the fourth corner.**

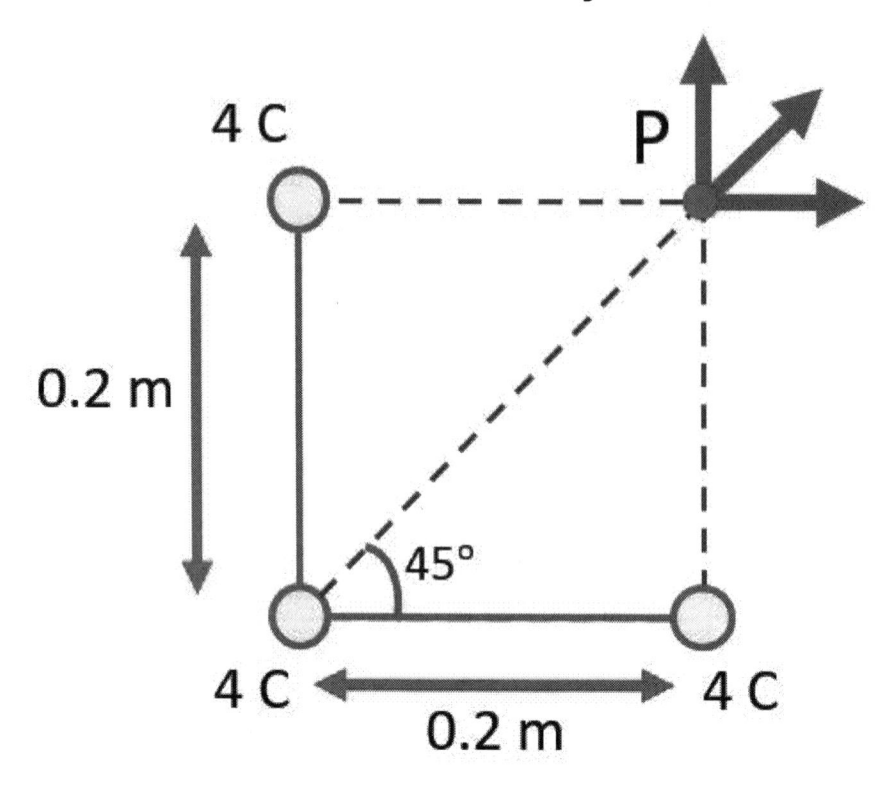

Assume the bottom left charge to be located at the origin, therefore the position vectors of charges Q_1, Q_2, Q_3 are,

$$\vec{r}_1 = 0.2\,\hat{i},\ \vec{r}_2 = 0,\ \vec{r}_3 = 0.2\,\hat{j}$$

Position vector of P: $\vec{r} = 0.2\,\hat{i} + 0.2\,\hat{j}$

$\vec{r} - \vec{r}_1 = 0.2\,\hat{j},\ |\vec{r} - \vec{r}_1| = 0.2$

$\vec{r} - \vec{r}_2 = 0.2\,\hat{i} + 0.2\,\hat{j},\ |\vec{r} - \vec{r}_2| = 0.2\sqrt{2}$

$\vec{r} - \vec{r}_3 = 0.2\,\hat{i},\ |\vec{r} - \vec{r}_3| = 0.2$

$$\vec{E}_1 = k \times 4 \times \frac{0.2\,\hat{j}}{0.2^3}$$

$$\vec{E}_2 = k \times 4 \times \frac{0.2\,\hat{i} + 0.2\,\hat{j}}{(0.2\sqrt{2})^3}$$

$$\vec{E}_3 = k \times 4 \times \frac{0.2\,\hat{i}}{0.2^3}$$

$$\vec{E} = \vec{E}_1 + \vec{E}_2 + \vec{E}_3$$
$$= \frac{4k}{0.2^2}\left(1 + \frac{1}{2\sqrt{2}}\right)(\hat{i} + \hat{j})$$

Gauss's Law:

The gauss's law states that "The electric flux passing through any closed surface is equal to the total charge enclosed by that surface"

$$\text{Flux } \phi = \oiint_S \vec{D}.\overrightarrow{dS}$$

10. **In a region D = x i + 2y j + 3z k, determine the flux radiating out of a sphere of 10 cm radius centered at the origin.**

$$\vec{D} = x\hat{i} + 2y\hat{j} + 3z\hat{k}$$
$$\nabla \cdot \vec{D} = 1 + 2 + 3 = 6$$

$$\text{Flux } \phi = \oiint_S \vec{D} \cdot \vec{dS}$$

Using Divergence Theorem,

$$\oiint_S \vec{D} \cdot \vec{dS} = \iiint_V \nabla \cdot \vec{D}\, dv$$
$$= 6 \iiint dv$$
$$= 6 \times \text{Volume of the sphere}$$
$$= 6 \times \frac{4}{3}\pi(0.1)^3$$
$$\phi = 0.025 \text{ C}$$

Electric Potential:

The electric potential is the amount of work needed to move a unit of electric charge from a reference point to the specific point in an electric field. It's a scalar quantity and its unit is Volt. The electric field intensity and the electric potential as related as,

$$\vec{E} = -\nabla V$$

11.

The potential at points in a plane is given by $V = a\cos\theta/r^2 + b/r$ where r and θ are the polar coordinates of a point in the plane, and a and b are constants. Find the electric field intensity E at any point.

$$V = \frac{a\cos\theta}{r^2} + \frac{b}{r}$$

$$\vec{E} = -\nabla V$$

$$\Rightarrow \vec{E} = -\left(\frac{\partial V}{\partial r}\hat{a}_r + \frac{1}{r}\frac{\partial V}{\partial \theta}\hat{a}_\theta\right)$$

$$\frac{\partial V}{\partial r} = -\frac{2a\cos\theta}{r^3} - \frac{b}{r^2}$$

$$\frac{\partial V}{\partial \theta} = -\frac{a\sin\theta}{r^2}$$

$$\therefore \vec{E} = \left(\frac{2a\cos\theta}{r^3} + \frac{b}{r^2}\right)\hat{a}_r + \left(\frac{a\sin\theta}{r^3}\right)\hat{a}_\theta$$

Magnetic Flux:

Magnetic flux is defined as the number of magnetic field lines passing through a given closed surface. It is a scalar quantity and its unit is Weber.

$$\phi = \vec{B}.\vec{A}$$

12. A square of side L meters lies in the x-y plane in a region where the magnetic field is given by B = B_0(2i + 3j + 4k) Tesla. Here B_0 is a constant. The magnitude of the flux passing through the square is?

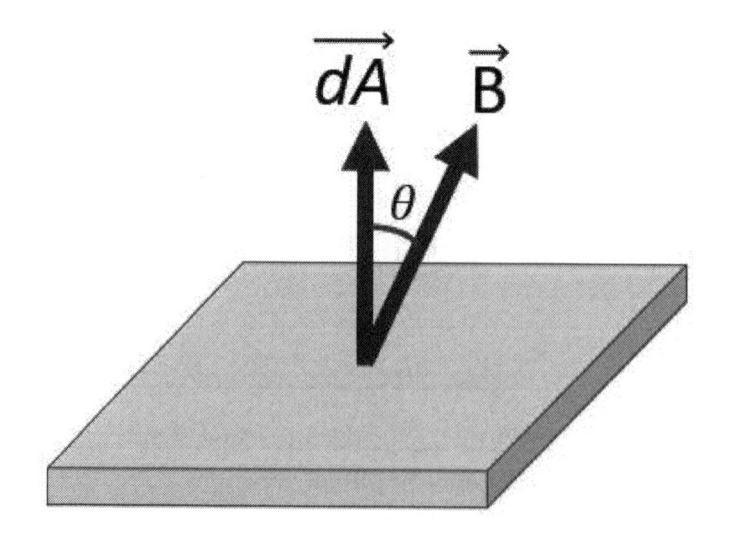

The square lies in the x-y plane, therefore the area vector points in the z direction,

$$\vec{A} = L^2 \hat{k}$$
$$\therefore \phi = \vec{B}.\vec{A}$$
$$= B_0(2\hat{i} + 3\hat{j} + 4\hat{k}).(L^2\hat{k})$$
$$= 4B_0L^2 \text{ Webers}$$

Lorentz Force:

The Lorentz Force is the force experienced by a moving charged particle due to electric and magnetic fields. The force experienced by a charge particle q moving with a velocity v through an electric field E and a Magnetic field B is given by,

$$\vec{F} = q\vec{v} \times \vec{B} + q\vec{E}$$

13. **A particle enters a uniform magnetic field and experiences an upward force as shown in the figure. Is the particle positively charged or negative charged?**

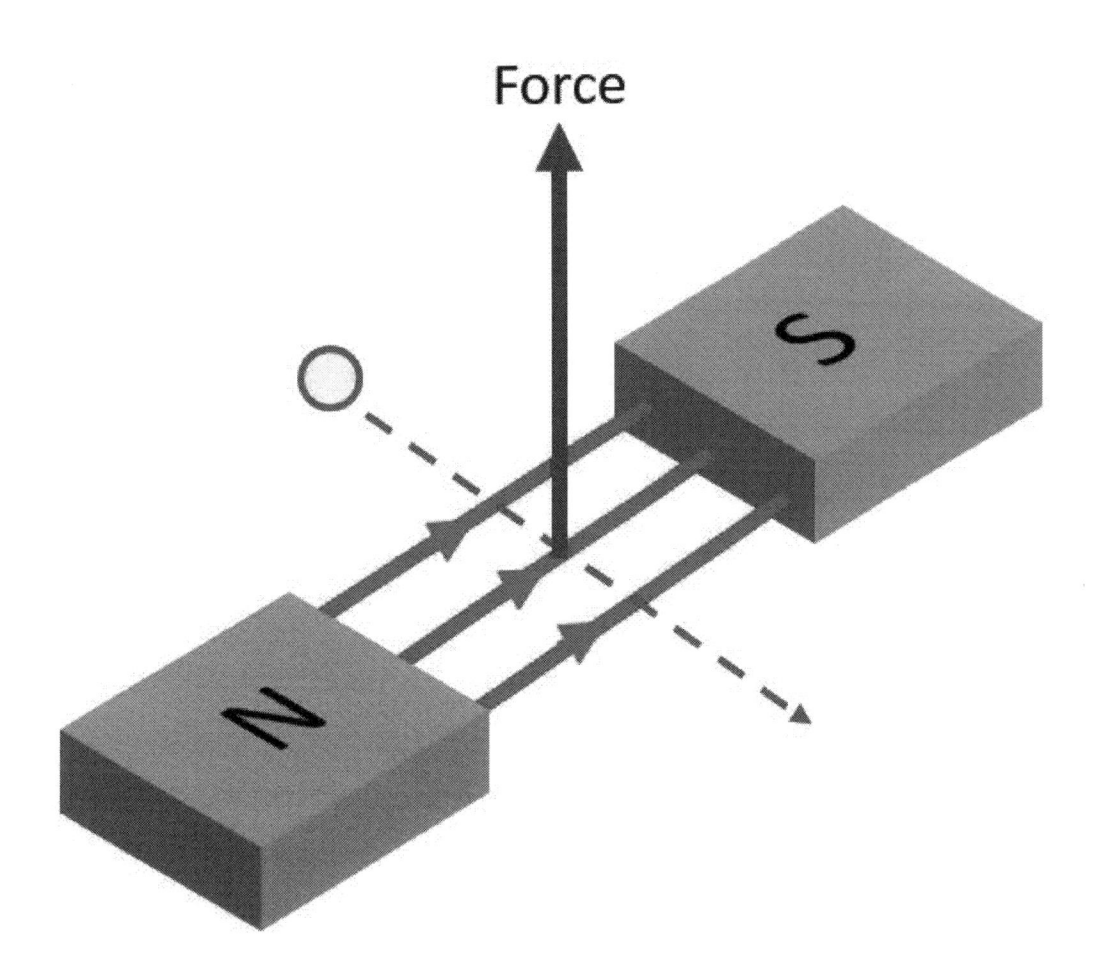

Applying the Flemings left hand rule we can see that the direction of the current is the same as the direction of motion of the charge. Which means that the charge is positive in nature.

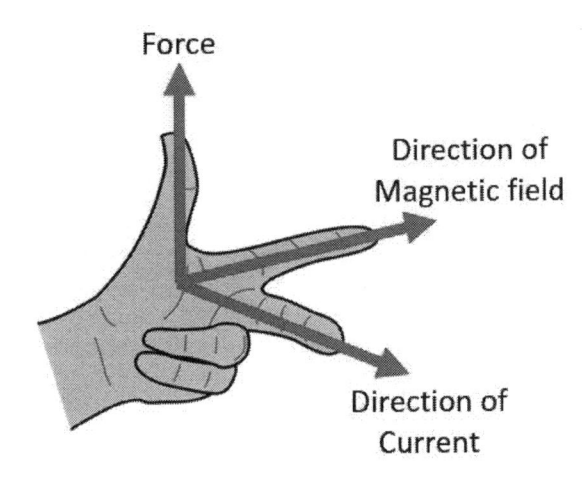

14. **An electron has a velocity of 10^6 m/s in the x-direction in a magnetic field B = 0.2i - 0.3j + 0.5k Wb/m². What is the**

electric field present if no net force is being applied to the electron?

$$Q = -1.6 \times 10^{-19} \, C,$$
$$\vec{v} = 10^6 \, \hat{i}$$
$$\vec{B} = 0.2\hat{i} - 0.3\hat{j} + 0.5\hat{k}$$

$$\vec{F} = Q(\vec{v} \times \vec{B}) + Q\vec{E} = 0$$

$$\therefore \vec{E} = - \begin{vmatrix} \hat{i} & \hat{j} & \hat{k} \\ 10^6 & 0 & 0 \\ 0.2 & -0.3 & 0.5 \end{vmatrix}$$

$$= 10^6 \, (0.5\hat{i} + 0.3\hat{k}) \, V/m$$

Maxwell's Equations:

Maxwell's equations are a set of 4 equations that form the foundation of classical electromagnetism.

Corresponding Law	Integral form	Point form
Gauss's Law for Electricity	$\oint_S \vec{D} \cdot \vec{dS} = \int_V \rho_v \, dv$	$\nabla \cdot \vec{D} = \rho_v$
Ampere-Maxwell Law	$\oint_C \vec{H} \cdot \vec{dl} = I_{enc} + \int_S \frac{\partial \vec{D}}{\partial t} \cdot \vec{dS}$	$\nabla \times \vec{H} = \vec{J} + \frac{\partial \vec{D}}{\partial t}$
Gauss's Law for Magnetism	$\oint \vec{B} \cdot \vec{dS} = 0$	$\nabla \cdot \vec{B} = 0$
Faraday's Law	$\oint \vec{E} \cdot \vec{dl} = -\frac{d}{dt}(\int_S \vec{B} \cdot \vec{dS})$	$\nabla \times \vec{E} = -\frac{d(\vec{B})}{dt}$

15.

Derive the Helmholtz equation $\nabla^2 E = \mu\varepsilon \, \partial^2 E / \partial t^2$ for lossless medium (J = 0) from the maxwell's equations. (Note D = εE, B = μH)

Maxwell's Equations for lossless medium are:

$$\nabla \times \vec{E} = -\mu \frac{\partial \vec{H}}{\partial t} \quad\quad\quad (1)$$

$$\nabla \times \vec{H} = \varepsilon \frac{\partial \vec{E}}{\partial t} \quad\quad\quad (2)$$

$$\nabla \cdot \vec{E} = 0 \quad\quad\quad (3)$$

$$\nabla \cdot \vec{H} = 0 \quad\quad\quad (4)$$

Taking curl on both sides of eqn1,

$$\nabla \times (\nabla \times \vec{E}) = -\mu \left(\nabla \times \frac{\partial \vec{H}}{\partial t}\right)$$

$$= -\mu \left(\frac{\partial (\nabla \times \vec{H})}{\partial t}\right)$$

Using vector identity,

$$\nabla \times (\nabla \times \vec{E}) = \nabla(\nabla \cdot \vec{E}) - \nabla^2 \vec{E}$$

$$\nabla^2 \vec{E} = \mu \left(\frac{\partial (\nabla \times \vec{H})}{\partial t}\right) \quad (\because \nabla \cdot \vec{E} = 0)$$

Substituting eqn2 in this equation,

$$\nabla^2 \vec{E} = \mu \left(\frac{\partial (\varepsilon \frac{\partial \vec{E}}{\partial t})}{\partial t}\right)$$

$$= \mu \varepsilon \frac{\partial^2 \vec{E}}{\partial t^2}$$

16. **Derive the expressions for the maxwell's laws in point form from the integral form. (Note D = εE, B = μH)**

Maxwell's Equations in Integral form are:

$$\oint_S \vec{D}.\vec{ds} = \int_V \rho_v \, dv \qquad \oint_S \vec{B}.\vec{dS} = 0$$

$$\oint_C \vec{H}.\vec{dl} = \int_S (\vec{J} + \tfrac{\partial \vec{D}}{\partial t}).\vec{ds} \qquad \oint \vec{E}.\vec{dl} = -\tfrac{d}{dt}(\int_S \vec{B}.\vec{dS})$$

Applying the Divergence theorem to eqns 1 and 2, we can convert the surface integrals to volume integrals as,

$$\int_V \nabla.\vec{D} = \int_V \rho_v \, dv \qquad \int_V \nabla.\vec{B} = 0$$

$$\boxed{\nabla.\vec{D} = \rho_v} \qquad \boxed{\nabla.\vec{B} = 0}$$

Similarly, applying the stokes theorem to eqns 3 and 4, we can convert the line integrals to surface integrals as,

$$\int_S \nabla \times \vec{H} = \int_S (\vec{J} + \tfrac{\partial \vec{D}}{\partial t}).\vec{ds} \qquad \int_S \nabla \times \vec{E} = -\tfrac{d}{dt}(\int_S \vec{B}.\vec{dS})$$

$$\boxed{\nabla \times \vec{H} = \vec{J} + \tfrac{\partial \vec{D}}{\partial t}} \qquad \boxed{\nabla \times \vec{H} = \tfrac{-\partial \vec{B}}{\partial t}}$$

7.3 FLUID MECHANICS

Fluid mechanics is the branch of physics concerned with the mechanics of fluids (liquids, gases, and plasmas) and the forces on them. It has applications in a wide range of disciplines, including mechanical, civil, geophysics, oceanography, meteorology etc.

Velocity Field:

The distribution of fluid velocities is mathematically described using velocity vector fields. The velocity at a point in the flow is a function of position and time(in case of unsteady flow).

17. Find the velocity vector at a point P(9,-2,1) of an unsteady velocity field V = (5xy² + t) i + (2z + 8) j + 18 k at instant t = 4. Also calculate its magnitude.

$$\vec{V} = (5xy^2 + t)\,\hat{\imath} + (2z + 8)\,\hat{\jmath} + 18\,\hat{k}$$

$$\vec{V}_P = (5(9)(-2)^2 + 4)\,\hat{\imath} + (2(1) + 8)\,\hat{\jmath} + 18\,\hat{k}$$
$$= 184\,\hat{\imath} + 10\,\hat{\jmath} + 18\,\hat{k}$$

$$|\vec{V}_P| = \sqrt{184^2 + 10^2 + 18^2}$$
$$= 185.1 \text{ m/s}$$

Vorticity:

Vorticity is a vector quantity that describes the microscopic local spinning motion in a fluid flow.

18. Find out if the velocity field V = -xy³ i + y³ j is irrotational or not.

$$\vec{V} = -xy^3\,\hat{\imath} + y^4\,\hat{\jmath}$$

$$\nabla \times \vec{V} = \begin{vmatrix} \hat{\imath} & \hat{\jmath} & \hat{k} \\ \dfrac{\partial}{\partial x} & \dfrac{\partial}{\partial y} & \dfrac{\partial}{\partial z} \\ -xy^3 & y^4 & 0 \end{vmatrix}$$

$$= \left(\dfrac{\partial(0)}{\partial y} - \dfrac{\partial(y^4)}{\partial z}\right)\hat{\imath} + \left(\dfrac{\partial(-xy^3)}{\partial z} - \dfrac{\partial(0)}{\partial x}\right)\hat{\jmath}$$
$$+ \left(\dfrac{\partial(y^4)}{\partial x} - \dfrac{\partial(-xy^3)}{\partial y}\right)\hat{k}$$

$$= 3xy^2\,\hat{k} \neq 0$$

The vorticity of the velocity field V is non zero, therefore the field is rotational in nature.

19. Find out the spin vector of the velocity field V = (x + y + z) i + (xy + yz + z²) j + (-3xz – z²/2 + 4) at a point P(1,1,1).

$$\vec{V} = (x^2 + y^2 + z^2)\,\hat{\imath} + (xy + yz + z^2)\,\hat{\jmath} + \left(-3xz - \frac{z^2}{2} + 4\right)\hat{k}$$

$$V_x = x^2 + y^2 + z^2$$
$$V_y = xy + yz + z^2$$
$$V_z = -3xz - \frac{z^2}{2} + 4$$

$$\vec{\omega} = \frac{1}{2}(\nabla \times \vec{V})$$

$$= \frac{1}{2}\begin{vmatrix} \hat{\imath} & \hat{\jmath} & \hat{k} \\ \dfrac{\partial}{\partial x} & \dfrac{\partial}{\partial y} & \dfrac{\partial}{\partial z} \\ V_x & V_y & V_z \end{vmatrix}$$

$$= \frac{1}{2}\left[\left(\frac{\partial(V_z)}{\partial y} - \frac{\partial(V_y)}{\partial z}\right)\hat{\imath} + \left(\frac{\partial(V_x)}{\partial z} - \frac{\partial(V_z)}{\partial x}\right)\hat{\jmath} + \left(\frac{\partial(V_y)}{\partial x} - \frac{\partial(V_x)}{\partial y}\right)\hat{k}\right]$$

$$= \frac{-(y+2z)}{2}\hat{\imath} + \frac{5z}{2}\hat{\jmath} - \frac{y}{2}\hat{k}$$

$$\vec{\omega}(1,1,1) = \frac{-3}{2}\hat{\imath} + \frac{5}{2}\hat{\jmath} - \frac{1}{2}\hat{k}$$

Acceleration Field:

The acceleration field describes the rate of change of velocities at different points in the fluid.

$$\frac{du}{dt} = \frac{\partial u}{\partial t} + u\frac{\partial u}{\partial x} + v\frac{\partial u}{\partial y} + w\frac{\partial u}{\partial z}$$

$$\frac{dv}{dt} = \frac{\partial v}{\partial t} + u\frac{\partial v}{\partial x} + v\frac{\partial v}{\partial y} + w\frac{\partial v}{\partial z}$$

$$\frac{dw}{dt} = \frac{\partial w}{\partial t} + u\frac{\partial w}{\partial x} + v\frac{\partial w}{\partial y} + w\frac{\partial w}{\partial z}$$

Where u,v and w are the x, y and z components of the velocity field.

20. A velocity field is given as V = 4tx $\hat{i}$ -2t²y $\hat{j}$ + 4xz $\hat{k}$, find the acceleration field.

$$u = 4tx, \quad v = -2t^2y, \quad w = 4xz$$

$$\frac{du}{dt} = \frac{\partial u}{\partial t} + u\frac{\partial u}{\partial x} + v\frac{\partial u}{\partial y} + w\frac{\partial u}{\partial z}$$
$$= 4x + 4tx(4t) = 4x + 16t^2x$$

$$\frac{dv}{dt} = \frac{\partial v}{\partial t} + u\frac{\partial v}{\partial x} + v\frac{\partial v}{\partial y} + w\frac{\partial v}{\partial z}$$
$$= -4ty - 2t^2y(-2t^2) = -4ty + 4t^4y$$

$$\frac{dw}{dt} = \frac{\partial w}{\partial t} + u\frac{\partial w}{\partial x} + v\frac{\partial w}{\partial y} + w\frac{\partial w}{\partial z}$$
$$= 4tx(4z) + 4xz(4x) = 16txz + 16x^2z$$

$$\therefore \vec{A} = (4x + 16t^2x)\,\hat{\imath} + (-4ty + 4t^4y)\,\hat{\jmath}$$
$$+ (16txz + 16x^2z)\,\hat{k}$$

APPENDIX

1. Triangle law of Vector Addition (Proof)

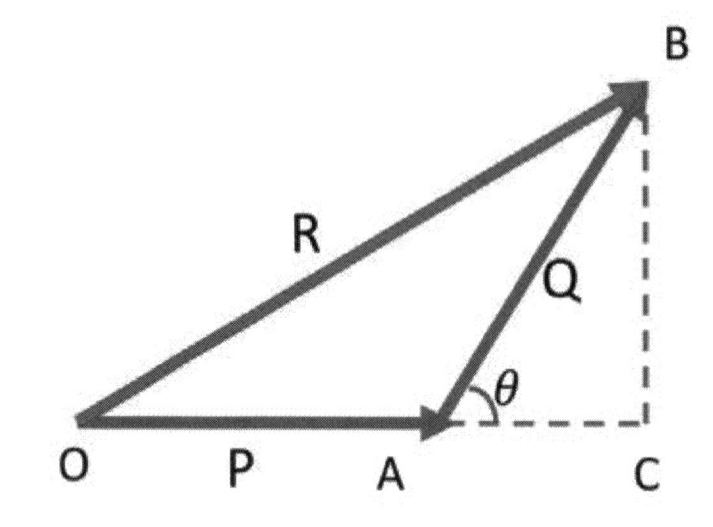

In $\triangle OCB$,
$$OB^2 = OC^2 + BC^2$$
$$= (OA + AC)^2 + BC^2$$

In $\triangle OCB$,
$$AC = AB \cos\theta$$
$$BC = AB \sin\theta$$

$\therefore OB^2 = (OA + AB \cos\theta)^2 + (AB \sin\theta)^2$
$= OA^2 + AB^2 \cos^2\theta$
$\quad + 2(OA)(AB) \cos\theta + AB^2 \sin^2\theta$
$= OA^2 + AB^2 + 2(OA)(AB) \cos\theta$

$$R = \sqrt{P^2 + Q^2 + 2PQ \cos\theta}$$

2. A x (B x C) = B(A.C)-C(A.B) (Proof)

Vector A x (B x C) lies in the plane of vectors B and C, therefore it can be expressed as the linear sum of vectors B and C.

$$A \times (B \times C) = mB + nC$$

Taking its dot product by A,
$$A \cdot (A \times (B \times C)) = m(A \cdot B) + n(A \cdot C)$$

Since $A \times (B \times C)$ is perpendicular to A,
$$A \cdot (A \times (B \times C)) = 0$$

$$\therefore m(A \cdot B) + n(A \cdot C) = 0$$
$$\Rightarrow \frac{m}{A \cdot C} = \frac{-n}{A \cdot B} = \alpha$$

$$\therefore A \times (B \times C) = \alpha((A \cdot C)B - (A \cdot B)C)$$

Let $A = \hat{i}$, $B = \hat{i}$, $C = \hat{j}$

$$A \times (B \times C) = \hat{i} \times (\hat{i} \times \hat{j}) = -\hat{j}$$
$$(A \cdot C)B = (\hat{i} \cdot \hat{j})\hat{i} = 0$$
$$(A \cdot B)C = (\hat{i} \cdot \hat{i})\hat{j} = \hat{j}$$

Substituting these values we can prove $\alpha = 1$

$$\therefore \boxed{A \times (B \times C) = (A \cdot C)B - (A \cdot B)C}$$

Printed in Great Britain
by Amazon